Germfree and Gnotobiotic Animal Models

Background and Applications

Bernard S. Wostmann, D. Sc.
Lobund Laboratory
University of Notre Dame
Notre Dame, Indiana

CRC Press
Taylor & Francis Group
Boca Raton London New York

CRC Press is an imprint of the
Taylor & Francis Group, an **informa** business

Acquiring Editor: Marsha Baker
Project Editor: Andrea Demby
Marketing Manager: Susie Carlisle
Direct Marketing Manager: Becky McEldowney
Cover design: Dawn Boyd
PrePress: Gary Bennett
Manufacturing: Sheri Schwartz

Library of Congress Cataloging-in-Publication Data

Wostmann, Bernard S.
    Germfree and gnotobiotic animal models : background and
applications / Bernard S. Wostmann.
        p. cm.
    Includes bibliographical references and index.
    ISBN 0-8493-4008-X (alk. paper)
    1. Laboratory animals.   2. Germfree animals.   3. Diseases--Animal
models.   I. Title.
        [DNLM: 1. Germ-Free Life.   2. Animals, Laboratory.   QW 56 W935g
1996]
QL55.W67   1996
591'.0724—dc20
DNLM/DLC
for Library of Congress   96-4752
                                                                                            CIP

FOREWORD

The desirability of having a truly germfree animal available for study in the biomedical field had been expressed by Louis Pasteur as early as the 1880s. Shortly thereafter, Nuttal and Thierfelder had been able to produce a germfree guinea pig, albeit short-lived. However, it was only in the late 1940s and early 1950s that it became possible to establish breeding colonies of germfree rats and mice. Although this opened the way for a multitude of studies in which that little-understood and uncontrollable factor, the "normal" microbial flora, could be eliminated, obviously the first question to be answered was "What kind of animal is this germfree animal?" This question was especially pertinent since Pasteur had not been certain that life under germfree conditions would even be possible.

However, life under germfree conditions proved to be possible. After colony production made available sufficient germfree rats and mice, and later guinea pigs and rabbits, the first two decades beginning in 1950 were largely devoted to the study of function and metabolism in the absence of a microbial flora. Yet even in those early days, after it was recognized that in the total absence of bacteria, e.g., dental caries did not develop, studies were started which used the germfree rat to introduce specific microorganisms to test their cariogenic potential and a beginning was made in the study of the effects of intestinal microorganisms on cholesterol and bile acid metabolism, obviously in support of the fight against cardiovascular disease.

Emphasis then gradually shifted to the application of the germfree and the gnotobiotic animal in studies of specific problems in the biomedical field; but in the meantime a younger generation of workers had taken over. Increasingly, they would need these animals in their studies; however, they might not be quite familiar with the specific characteristics and anomalies of this animal model, which could considerably affect the interpretation of their experimental data. It is for them that we have tried to bring the most salient points of the earlier and more recent studies together without going into too much detail, but giving references to the pertinent literature as often as possible. We have also tried to interpret earlier data against results of more recent studies.

Although the majority of the studies with germfree and gnotobiotic animals have been carried out with rats and mice, the first animal to be obtained germfree was the guinea pig, soon to be followed by the germfree chicken. In both cases the fact that the newborn animal could survive without maternal care explains these choices, whereas in the case of the chicken the possibility of external sterilization of the egg added to its attractiveness. Later experience with the difficulties of hand-feeding newborn Cesarian-derived rats and, particularly, newborn mice, would fully justify these choices. This also explains why early studies, when absence or control of the microflora was sought, were carried out with "first generation" germfree chickens and guinea pigs. Then, once colony production of germfree rats and mice had been established, which had necessitated extensive experience with the hand-feeding of newborns, the germfree rabbit seemed the next logical choice at the time because of its importance for immunological studies. Although hand-feeding of the newborn Cesarian-derived germfree rabbit posed no exceptional difficulties, problems posed by its very much distended cecum, in addition to unexpected nutritional problems, took time to solve. By the time adequate solutions had been found, the emphasis of most immunological studies had shifted to the use of mice, and the germfree rabbit became an almost forgotten research model. To an extent, the potential role of the germfree rabbit in immunological studies has been taken over by the germfree and gnotobiotic piglet in the extensive studies of Dr. Yoon Kim and co-workers at the Department of Microbiology and Immunology of the Chicago Medical School. Of particular interest is the fact that they were able to rear newborn germfree piglets on a diet of very low antigenicity.

The germfree gerbil has been of special interest to this author because, as described in Chapters II and IV, its cholesterol and bile acid metabolism are so close to those of the human. This model, especially as a controlled polyassociate, may be of great help in solving problems related to cardiovascular disease.

The GF dog, developed under the auspices of Dr. Charles Yale and Dr. Jim Heneghan at the LSU School of Medicine, has served as an important tool in the study of the pathology of intestinal strangulation. Another line of research led to the production of germfree and gnotobiotic pigs at the University of Guelph, until recently under the supervision of Dr. Paul Miniats. Here the emphasis was on the production of a pathogen-free animal which, as such, could be introduced into the swine population for commercial purposes. Soon gnotobiotic pigs would also be used extensively to study the pathology of the bacterial diseases that plague commercial pork production (see Chapter XI). Last but not least, the great amount of work which went into the various endeavors to save children born with severe combined immune deficiency (SCID) should be mentioned. In this country this led to the story of "The Boy in the Bubble", which is described in the Proceedings of the VIII International Symposium on Germfree Research held at the University of Notre Dame in 1984.

This monograph has been limited mainly to germfree and gnotobiotic rats, mice, and gerbils, all of which we have been closely associated with. As mentioned earlier, the vast majority of the more critical studies have been carried out with these animal models. Whenever pertinent, results obtained with germfree and/or gnotobiotic chickens, guinea pigs, rabbits, dogs, and pigs have been introduced. Chapters I to IX concentrate mostly on the anomalies which may occur as a result of the germfree and of certain gnotobiotic states. Chapters X, XI, and XII describe the application of these animal models to a variety of problems as they occur in the biomedical sciences.

We are acutely aware of the fact that some work that should have been included may have been overlooked. For this we apologize; no personal files are ever complete. For completeness we often had to rely on abstracts from *Index Medicus*. Our sincere thanks are due to Dr. Morris Pollard and Dr. Paul Weinstein, who were willing to give much needed help in fields we were least familiar with.

THE AUTHOR

Bernard S. Wostmann, D.Sc., received a M.Sc degree in biological chemistry from the University of Amsterdam, The Netherlands in 1945. He received his D.Sc degree in 1948 while working in the Department of Physiological Chemistry of the Medical School of Amsterdam University, where he joined the staff as a lecturer in Physiological Chemistry. In 1950 to 1951 he was a Rockefeller Research Fellow at the California Institute of Technology, working under Dr. Linus Pauling and Dr. Dan H. Campbell on antigen-antibody combining ratios. In 1951 to 1955 he served as Assistant Professor in above department, while also being a member of the staff of The Netherlands Institute of Nutrition. In 1955 he became Assistant Research Professor at the Lobund Laboratory of the University of Notre Dame in charge of its Division of Biochemistry and Nutrition. In 1958 he became an Associate and from 1965 to 1988, Professor, first in the Department of Microbiology, later in the Department of Biological Sciences. He taught Gnotobiology, Nutrition, and the chemical background of antibiotics to undergraduate and graduate students. In 1988 he retired as an active member of the department, while retaining his connection with the Lobund Laboratory. His research has been supported by grants from the National Institutes of Health, the National Science Foundation, the Nutrition Foundation, the University of Notre Dame, Erasmus University, Rotterdam, The Netherlands, and various smaller foundations.

He has written approximately 190 scientific papers, has edited one book, and has lectured in the U.S., Canada, Mexico, Brazil, Europe, Japan, and China. Memberships in scientific societies include the American Association for the Advancement of Science, the Association for Gnotobiotics, the International Association for Gnotobiotics, the American Institute of Nutrition, the Society for Experimental Biology and Medicine, the Indiana Academy of Science, the Society of Sigma Xi, and the International Committee for Laboratory Animal Science (ICLAS).

TABLE OF CONTENTS

Chapter I

INTRODUCTION

Since the 1950s, when colony production of germfree (GF)* rats and GF mice was established, the use of these animal models in the biomedical sciences has become widespread. The models have been used to attack a wide variety of problems, from the bacterial activation of carcinogens to the transplantation of allogeneic bone marrow, and from cholesterol metabolism to the effects of dietary restriction on the aging syndrome. However, not always does the investigator realize that in a number of aspects the GF animal may be quite different from his or her familiar animal model with only the microbial factor removed. Cecal enlargement in GF Rodentiae and Cuniculae, and the underdeveloped immune system of the GF animal in general, are factors that are recognized by most investigators. However, the differences go beyond this, and appear to affect many aspects of function and metabolism: metabolic rate, gastrointestinal function, specific and quantitative aspects of immune function, and the many aspects of biochemical homeostasis in which, because of involvement of the enterohepatic circulation, the bacterial factor otherwise would have played a role. Thus, while on the one hand the GF animal is an ideal model to study the many aspects of one life form in the absence of all other forms of life, this absence of all other (microbial) life forms gives rise to a difference in homeostasis which cannot be ignored by the investigator.

---

* The nomenclature used in this publication will be limited to terms of general usage. **Gnotobiotic**, derived from the Greek "gnotos" meaning known, and "biota", together, indicate an animal with associated flora or fauna that is fully defined by accepted current methodology. **Germfree** (axenic) indicates a gnotobiote which is free of all demonstrable associated forms of life including bacteria, viruses, fungi, protozoa, and other saprophytic or parasitic forms. **Defined Flora** indicates a gnotobiote maintained under isolator conditions which has been intentionally associated with one or more defined microbial species. **Conventional** (controls) indicates animals of the same genetic background, maintained on the same sterilized diet, but in an open environment. A restriction to this are **Isolator Conventional** animals, which are maintained under isolation most or all of the time (see Chapter II). **Conventionalized** animals are ex-germfree animals which have been associated with the flora of conventional controls, but do not necessarily duplicate the microflora of these animals in every detail. Abbreviations used: Gnotobiotic: GN; Germfree: GF; Conventional: CV.

When reviewing the field of gnotobiology the first thing to remember is that the potential importance of the GF animal was first recognized more than 100 years ago. In 1885 Louis Pasteur wrote:

> For several years during discussions with young scientists in my laboratory, I have spoken of the interest in feeding a young animal from birth with pure nutritive material. By this expression I mean nutritive products which have been artificially and totally deprived of common microorganisms.
>
> Without affirming anything, I do not conceal the fact that if I had the time, I would undertake such a study, with the preconceived idea that under such conditions life would have become impossible.
>
> If this work could be developed simply, one could then consider the study of digestion with the systematic addition to the pure food, of one or another single microorganism, or diverse microorganisms with well defined relationships. (Pasteur, L., *C. R. Acad. Sci. Paris*, 100, 68, 1885.)

Thus, at a time when nutrition started to become a significant issue, and the importance of bacteria had just been recognized, the need for a more defined animal model free of microbial associates was already obvious to Pasteur. Despite his doubt about the feasibility of developing such a model, not more than 10 years elapsed before Nuttal and Thierfelder, working at the University of Berlin, produced the first GF animal, a GF guinea pig.[1]

However, it took more than 50 years from the time the first GF guinea pig was obtained (around 1895) until the first breeding rat colonies were successfully established (in the late 1940s).[2,3] This was caused largely by a lack of knowledge about nutrition. The years between 1900 and 1950 saw major developments in the science of nutrition, and the development of the GF animal model largely parallels these endeavors. It is therefore pertinent that we look for a moment at the history of the GF animal.

As mentioned earlier, Nuttal and Thierfelder obtained the first GF guinea pigs less than 10 years after Pasteur had suggested such an experiment with "...the preconceived idea that life under those conditions would have become impossible." Germfree life, however, did prove to be possible, but only for a short time, since nutritional requirements obviously were not met after the harsh sterilization of the dietary materials. In this case the longest surviving guinea pig lived only 13 days. Here, it should be remembered that dietary deficiency (as the cause of beriberi) had been recognized only in the 1890s,[4] and that the concept of vitamins was introduced by Funk only in 1912.[5]

Later work with GF chickens, which were fed what at that time was thought to be a diet containing every possible requirement, nonetheless never achieved survival beyond 6 weeks. At that age the animals were clearly in poor condition. However, when a contamination with *E. coli* occurred, survival was extended to about 15 weeks. Conceivably, survival in the GF state was possible but hitherto had been limited by deficient nutrition, whereas the intestinal bacteria appeared to be able to provide one or more necessary but as yet unknown nutrients.[6]

Although the work with GF chickens had been carried out in Metchnikoff's laboratory in Paris, the importance of the GF animal was also recognized in Germany. In 1908 Küster reported raising a GF goat on a fortified milk diet. The animal remained germfree for 6 weeks, when it became contaminated. His was an extremely well-conducted and well-documented study, indicating that mammals could develop normally under GF conditions.[7] Shortly thereafter, World War I brought these studies to an end, both in France and in Germany. It was only after World War II that they resumed in those countries.

Sweden, however, had remained neutral, and it was there that GF studies continued. In 1932 Gösta Glimstedt, Professor of Pathology at the University of Lund, reported a more successful production of GF guinea pigs.[8,9] With improved nutrition, he managed to extend survival to 2 months. He was able to obtain numerous guinea pigs in the GF state, which were then sacrificed for scientific observation while obviously in good health. Glimstedt was especially interested in the interaction of the microflora with the lymphoid tissue, which he found to be quite underdeveloped in the GF state. His precise description of the equipment shows isolators of almost modern sophistication.[9] When we visited Lund in 1958, studies with GF animals, now mostly rats, were continuing under the direction of Dr. Bengt E. Gustafsson.

In the meantime, unaware of the work in Sweden, James A. (Art) Reyniers at the University of Notre Dame had started work on GF animals with a similar concept in mind: development of an animal model functioning without the interplay of uncontrolled but viable associates. Working since 1928 on the development of the proper equipment, in 1946 he reported on the rearing of GF albino rats. This involved Cesarian derivation of the pups at term, with subsequent around-the-clock feeding of a milk formula via a rubber nipple attached to a syringe. Both Reyniers and Gustafsson reported the first successful establishment of a breeding rat colony in the late 1940s,[2,3] followed later by a similar announcement by Miyakawa et al. at the University of Nagoya, Japan.[10] Then in 1952 Dr. Julian Pleasants was able to establish, via Cesarian derivation and hand-feeding, a colony of Swiss-Webster albino mice at the Lobund Laboratory of the University of Notre Dame.[11] In 1966 Gordon[12] published a short but comprehensive paper describing the characteristics of the then available GF rats and GF mice, with special emphasis on their potential for gerontological studies.

Because of the comparative ease with which GF chickens could be obtained, they were extensively used in the 1950s and 1960s. Some of the early work at the Lobund Laboratory used this GF model. More extensive studies were done by Dr. Marie Coates and associates at the National Institute for Research in Dairying at the University of Reading, England.[13,14]

However, since the mature birds require much space and extensive care, this model was largely abandoned in the 1970s. During those years many other species were obtained "germfree", and breeding colonies of GF rabbits and GF guinea pigs were established.[15,16] In 1978 Wostmann *et al.*[17] reported having obtained the GF gerbil by foster nursing of Cesarian-derived pups on GF mice. These animals were subsequently hexa-associated to reduce an overly enlarged cecum. The resulting gnotobiotes were then able to breed, and proved especially valuable in cholesterol and bile acid research (see Chapters III and IV).

The above clearly indicates that although sophisticated equipment was available in the 1930s, it took another 20 to 30 years until enough was known about nutrition to establish the correct dietary formulations to successfully hand-feed Cesarian-derived rats and rabbits whenever necessary, and to maintain the weaned animals in good health during a normal life span. However, derivation of GF mice always met with severe problems due to the technical difficulty of hand-feeding the newborn pups sufficient diet to ensure survival. Most, if not all GF mice presently available originate via cross-suckling from the GF Swiss-Webster mice originally obtained by Pleasants.[11]

Many GF species have been studied, and not all under ideal conditions. Only for those species for which nutritional requirements are known in great detail can we now compose a diet which, after the necessary sterilization, will definitely cover those requirements adequately. With this restriction in mind, it can be said that GF animals appear to grow as well as their CV counterparts. Whereas reproduction may be hampered in GF rats and GF mice by cecal enlargement, it proves to be quite adequate, although generally not as good as under conventional conditions.[18] Whenever adequate conditions prevail, these GF animals appear to be in good health and to live longer than their conventional (CV)* counterparts[19,20] (see Chapter II).

How do we define, at present, the "germfree state"? Presumably the old definition, given by Dr. Morris Wagner, still holds:

> A germfree animal is free of bacteria, yeasts, molds, viruses, protozoa, parasites and all other recognizable life forms.[21]

Thus far quite a number of animal species have been derived "germfree", but as far as is known the only animal answering to the above definition is the GF rat. The Wistar and Sprague-Dawly rats maintained at the Lobund Laboratory are, to the best of our knowledge, really germfree. This includes freedom from viruses by the standards described by Kajima and Pollard,[22] and also absence of mycoplasma[23] and *Pneumocystis carinii*.[24] This statement was later extended to the Fischer and Buffalo rats maintained at the Lobund Laboratory. We hesitate to extend it to all GF

---

* See footnote page 1.

rats, because not all strains have been properly tested. Also, several years ago it was found that almost all commercially produced "germfree" rats in the U.S. carried *P. carinii*.[25] To our knowledge, no specific statement has been made for GF rats produced in Europe or Japan. We know that our, and presumably all, GF mice carry leukemia virus,[26] whereas GF dogs have a tendency to carry parasites.[27] Of most other so-called "germfree" animals only the bacterial status has been established beyond reasonable doubt.

It is not within the scope of this monograph to go into details of the technology of the production of germfree animals. For this the reader is referred to Coates and Gustafsson's *The Germfree Animal in Biomedical Research*[28] and the various Proceedings of the International Symposia of Gnotobiology.[29-33] It should be mentioned, however, that due to the relatively small number of pregnant CV females used to obtain GF young by Cesarian derivation, genetic considerations are important. To maintain close genetic proximity between GF and CV rat and mouse colonies, the present CV colonies are usually derived by conventionalization of GF stock, with further reintroduction of GF males and females at specific intervals to maintain genetic proximity. For a further elaboration of this problem see the paper by Rapp and Burow.[34]

In the following pages reference will be made regularly to the Lobund Aging Study. This study was carried out in the 1980s and contains a large body of data on GF and CV male Lobund-Wistar rats. Since not all of this material has been published in a form suitable for this monograph, any material taken from this study will be referred to as originating from the "Lobund Aging Study", with proper reference to the published material wherever possible.*

At the present time GF rats and GF mice are commercially available, although the above-mentioned restrictions of the term "germfree" have to be taken into account. This allows for the following broad fields of study:

---

* The Lobund Aging Study was designed to study the effect of a 30% reduction in food intake on the aging process in GF and CV rats. It consisted of two experimentally and two *ad libitum*-fed control groups, each gradually expanded to include between 120 and 150 male Lobund-Wistar rats. Material was obtained from a number of age groups between 2 and 45 months. Although most of the analytical work was done at the Lobund Laboratory, blood analysis of general application was done on the Technicon SMASC (Computerized Sequential Multiple Analyzer) at the South Bend Medical Foundation. Also, part of the material harvested from these rats was made available to other laboratories with special expertise in certain fields. Since not all data derived from this study have (as yet) been published, such material will be referred to as "Lobund Aging Study, unpublished material", with name recognition wherever possible. Except for certain autopsy data, the data used in this publication refer only to material harvested from animals sacrificed in obviously healthy condition.The Lobund Aging Study was performed under the direction of Dr. Morris Pollard, director of the Lobund Laboratory, University of Notre Dame in the able hands of Dr. David L. Snyder.

1. Study of the anatomy, physiology, and pathology of the genotype in the absence of other life forms. Among other things, this makes "clean endpoint studies" possible in the field of gerontology, where decline and death can now be studied without the otherwise unavoidable interference of pathogenic microorganisms.

2. Evaluation of the role of the microflora in function and metabolism of the host.

3. Study of the effect of specific microorganisms on the host, as exemplified by the work on dental caries, and on the ontology of the immune response.

4. Controlled studies of the interaction of specific microorganisms within the host.

5. Controlled studies of the action of chemicals (e.g., carcinogens) and/or biological materials (e.g., LPS) on the host, and of the action of the microflora on these processes.

REFERENCES

1. Nuttal, G. H. F. und Thierfelder, H., Thierishes Leben ohne Bacterien im Verdauungskanal, *Z. Physiol. Chem.*, 21, 109, 1895-1896.
2. Reyniers, J. A., Trexler, P. C., and Ervin, R. F., Rearing germfree albino rats, *Lobund Report #1*, Notre Dame University Press, Notre Dame, IN, 1946.
3. Gustafsson, B. E., Germfree rearing of rats: general technique, *Acta Pathol. Microbiol. Scand. Suppl.*, 73, 1, 1948.
4. Eijkman, C., Report of the investigations carried out in the Laboratory of Pathology and Bacteriology, Weltevreden, *Med. J. Dutch East Indies*, 30, 294, 1890.
5. Funk, C., On the chemical nature of the substances which cure polyneuritis in birds induced by a diet of polished rice, *J. Physiol.*, 53, 395, 1911-1912.
6. Cohendy, M., Expérience sur la vie sans microbes, *Ann. Inst. Pasteur*, 26, 106, 1912.
7. Küster, E., Die Gewinnung, Haltung und Aufzucht keimfreier Tiere und ihre Bedeutung für die Erforschung natürlicher Lebensvorgänge, *Arb. Kais. Gesundh. Amtes*, 48, 1, 1915.
8. Glimstedt, G., Das Leben ohne Bakterien. Sterile Aufziehung von Meerschweinchen, *Anat. Anz.*, 75, 79, 1932.
9. Glimstedt, G., Der Stoffwechsel bacterienfreier Tiere. I. Allgemeine Methodik, *Skand. Arch. Physiol.*, 74, 48, 1936.
10. Miyakawa, M., Studies of rearing germfree rats, in *Advances in Germfree Research and Gnotobiology*, Miyakawa, M. and Luckey, T.D., Eds., CRC Press, Boca Raton, FL, 1967, 48.
11. Pleasants, J. R., Rearing germfree cesarean-born rats, mice, and rabbits through weaning, *Ann. N.Y. Acad. Sci.*, 78, 116, 1959.
12. Gordon, H. A., Germfree animals in the study of microbial effects on aging, in *Perspectives in Experimental Gerontology*, Charles C Thomas, Springfield, IL, 1966, 295.
13. Coates, M. E., Animal production and rearing: Chickens and quail, in *The Germfree Animal in Research*, Coates, M.E., Ed., Academic Press, London, 1968, 79.
14. Coates, M. E., Production of germfree animals: Birds, in *The Germfree Animal in Biomedical Research*, Coates, M.E. and Gustafsson, B. E., Eds., Laboratory Animals Handbooks 9, Laboratory Animals Ltd., London, 1984, 79.

15. Reddy, B. S., Pleasants, J. R., Zimmerman, D. R., and Wostmann, B. S., Iron and copper utilization in rabbits as affected by diet and germfree status, *J. Nutr.*, 87, 189, 1965.

16. Pleasants, J. R., Reddy, B. S., Zimmerman, D. R., Bruckner-Kardoss, E., and Wostmann, B. S., Growth, reproduction and morphology of naturally-born, normally suckled germfree guinea pigs, *Z. Versuchstierk.*, 9, 195, 1967.

17. Wostmann, B. S., Beaver, M., Bartizal, K., and Madsen, D., Gnotobiotic gerbils, in Proc. IVth Int. Symp. Contamination Control, Vol. 4, Washington, D.C., 1978, 132.

18. Saito, M. and Nomura, T., Production of germfree animals. I. Small mammals, in *The Germfree Animal in Biomedical Research*, Coates, M.E. and Gustafsson, B.E., Eds., Laboratory Animals Ltd., London, 1984, 33.

19. Pollard, M. and Wostmann, B. S., Aging in germfree rats: The relationship to the environment, diseases of endogenous origin, and to dietary modification, in: *8th Symposium of the International Council for Laboratory Animal Science (ICLAS), Vancouver*, Archibald, J., Ditchfield, J., and Rowsell, H.C., Eds., Gustav Fischer, New York, 1985, 181.

20. Gordon, H. A., Bruckner-Kardoss, E., and Wostmann, B. S., Aging in germfree mice: Life tables and lesions observed at natural death, *J. Gerontol.*, 21, 380, 1966.

21. Wagner, M., Determination of germfree status, *Ann. N.Y. Acad. Sci.*, 78, 89, 1959.

22. Kajima, M. and Pollard, M., Distribution of tumor viruses in germfree rodents, in *Sixth International Congress for Electron Microscopy*, Vol. II, Uyeda, R., Ed., Maruzen Co. Ltd. Tokyo, 1966, 215.

23. Lutsky, I and Organick, A., Recovery of mycoplasma from gnotobiotic systems, *Lab. Anim. Care*, 18, 610, 1968.

24. Wagner, M., Absence of *Pneumocystis carinii* in Lobund germfree and conventional rat colonies, *Prog. Clin. Biol. Res.*, 181, 55, 1985.

25. Pifer, L. L., Lattuada, C. P., Edwards, C. C., Woods, D. R., and Owens, D. R., *Pneumocystis carinii* infection in germfree rats: implications for human patients, *Diagn. Microbiol. Infect. Dis.*, 2, 23, 1984.

26. Pollard, M., The viral status of "germfree mice", *Natl. Cancer Inst. Monogr.*, 20, 167, 1966.

27. Gibson, J. P., Griesemer, R. A., and Koestner, A., Experimental distemper in gnotobiotic dogs, *Pathol. Vet.*, 2, 1, 1965.

28. Coates, M. E. and Gustafsson, B. E., Eds., *The Germfree Animal in Biomedical Research*, Laboratory Animal Handbooks, Vol. 9., Laboratory Animals Ltd., London, 1984.

29. Heneghan, J. B., Ed., Germfree Research. The Biological Effects of Germfree Environment, *Proceedings of the IV International Symposium on Germfree Research*, Academic Press, New York, 1973.

30. Sasaki, S., Ozawa, A., and Hashimoto, K., Eds., Recent Advances in Germfree Research, *Proceedings of the VII International Symposium on Gnotobiology*, Tokyo University Press, Tokyo, 1981.

31. Wostmann, B. S., Ed., Germfree Research. Microflora Control and its Application to the Biomedical Sciences, *Proceedings of the VIII International Symposium on Germfree Research*, Alan R. Liss, New York, 1985.

32. Gnotobiology and Its Applications, *Proceedings of the IX International Symposium on Gnotobiology*, Ed. Fond. Marcel Mérieux, Lyon, France, 1987.

33. Heidt, P. J. et al., Eds., Experimental and Clinical Gnotobiology, Proc. Xth Int. Symp. Gnotobiology, *Microecol. Ther.*, 20, 1, 1990.

34. Rapp, K. G. and Burow, K., Genetic problems due to small numbers of outbred animals within an isolator, in *Clinical and Experimental Gnotobiotics. Proceedings of the VI International Symposium on Gnotobiology*, Fliedner, T. et al., Eds., Gustav Fischer, New York, 1979, 73.

Chapter II

EARLY GROWTH, BODY WEIGHT, REPRODUCTION, AND LIFE SPAN

GENERAL ASPECTS

The above factors, when carefully monitored, will always be among the best indicators of the well-being of an animal under the conditions in which it is maintained. This is especially true for animals reared in the GF state, since subnormal performance will suggest deficiencies in diet, housekeeping, or both. Of the long-term studies, only those done in the 1970s and 1980s in which diets with little or no casein were used can be judged entirely reliable, because of casein's nephrosis-inducing potential.[1] In earlier work with GF mice, nephrosis proved to be a major cause of death in aging GF animals.[2] Also, in studies using GF and GN animals, the effect of the considerable antigenic load of sterilization-killed bacteria in casein may have to be taken into account (see Chapter IV).

GROWTH

As far as the presently available data indicate, early growth of GF and CV animals appears comparable whenever nutritional adequacy of food after sterilization is ensured. This has been definitely established in GF rats[3] and GF mice.[4] The same can be said for GF beagles, once they are beyond the mandatory phase of hand-feeding.[5] Even the early experiment of Küster, who obtained a GF goat as early as 1908, indicated comparable growth.[6] Some indication exists that the early growth of GF chickens might be slightly better than that of chickens kept under the usual laboratory conditions,[7] but to our knowledge no such data exist for the rodents under consideration.

BODY WEIGHT OF THE MATURE ANIMAL

Except for the aforementioned study by Gordon *et al.*,[2] the only large body of data on GF rodents is that of the Lobund Aging Study (see Chapter I), in which well over 100 GF and over 100 CV male Lobund-Wistar (L-W) rats were followed over their entire life span. These animals were maintained on natural ingredient colony diet L-485 (see Chapter V, Table 2), in use at the Lobund Laboratory since 1968. The data show that starting at approximately 6 months of age the usual increase in body weight is somewhat less in the GF than in the CV rats. The GF L-W rats stop gaining weight after about 1 year, while their CV counterparts continue to gain some weight until about 18 months of age (Figure 1).[3] In mice the presently available data do not seem to indicate such a difference. The statistical significance of the difference in body weight between the GF and CV adult rats became obvious only because of the large numbers of animals in the above-mentioned study. The difference obviously increases when body weights corrected for cecal weights are considered, since the ceca of the GF rodents are enlarged by a factor 4 to 8, depending on diet.[8]

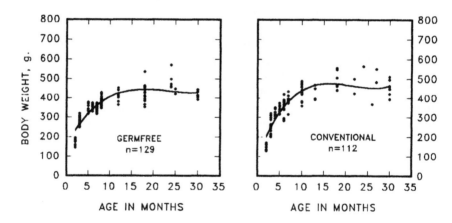

FIGURE 1
Body weights of male germfree and conventional Lobund-Wistar rats. At 18 months of age germfree rats (n = 17) weighed 430 g, conventional rats (n = 13) 471 g. $p = 0.01$.

The data available at this time for the GF gerbil only allow the statement that growth appears to be in the low to normal range, presumably because of its greatly enlarged cecum.[9]

REPRODUCTION

Provided that their nutritional requirements are met, GF rats and GF mice reproduce adequately. The females appear to remain productive as long as their CV counterparts, although the impression prevails that the number of young born during their reproductive life is somewhat less than in the CV animals.[10] Umehara et al.[11] noticed a decline in copulation and implantation ratios in GF ICR mice and speculated that the absence of an intestinal microflora would affect the spectrum of the steroid hormones that regulate reproduction. Gustafsson et al. have shown the extensive effect of the intestinal microflora on steroid hormones in the enterohepatic circulation.[12-14]

Another factor potentially affecting the reproductive capability of GF rats and GF mice may be the aforementioned cecal enlargement (for details see Chapter III). Maintained on an adequate diet, GF L-W rats show an approximately fivefold enlargement of the cecum.[15] GF gerbils, which show an eight- to tenfold enlargement, never reproduced in our laboratory. Only after association with a defined microflora consisting of Lactobacillus brevis, Streptococcus faecalis, Enterobacter aerogenes, Staphylococcus epidermidis, Bacteroides fragilis var. vulgatus, and a Fusobacterium sp., which brought cecal size down to about 1/3 of its GF size, did the animals reproduce.[9] The effect of the cecum was also quite conspicuous in the GF rabbit, which required almost total cecectomy for the first reproduction to occur, although later modifications of the diet made normal reproduction possible.[16] Obviously, cecal enlargement is a factor inherent to GF Rodentiae and Cuniculae that may affect reproductive capability.

GF guinea pigs at one time were colony-produced at the Lobund Laboratory.[17] Whereas Cesarian-born GF guinea pigs generally did not reproduce during the first year of life, the colony-produced animals, which had been housed in harem groups, reproduced at the normal age of about 4 months. Apparently the guinea pigs reared in isolation after Cesarian birth needed time to learn efficient mating procedures.[18] The colony-produced young grew at a somewhat smaller rate, presumably due to the stress of the very enlarged cecum, which also seemed to affect the number of young born.

LIFE SPAN

GF rats and GF mice appear to live longer than their CV counterparts raised under the usual animal house conditions, even though we have to consider the cecal enlargement inherent to the GF state as a potential negative factor. Especially in older animals, the heavy cecum occasionally leads to intestinal twists and to strangulation,[8] but notwithstanding the potential for this lethal occurrence, at the Lobund Laboratory both GF mice and GF rats live longer than their CV counterparts.[2,19]

Although the early study by Gordon et al.[2] may have been flawed to an extent by the fact that the diet used at the time (L-462,[20] for details see Chapter V) contained casein and other milk proteins, which recently have been shown to cause nephrosis later in life,[1] the difference in life span of the two groups of mice was substantial (Table 1). The GF and CV Swiss-Webster (S-W) mice used in this study had originally been derived from mice obtained from Harlan Farms, Cumberland, IN. As a customary part of the Lobund Laboratory management procedures, GF mice derived from this strain had been reintroduced at regular intervals into the CV colony to insure close genetic proximity.

TABLE 1
Lifespan of Germfree and Conventional Swiss-Webster Mice Maintained on Diet L-462[20]

	Germfree	Conventional
Males	23.6 ± 0.5 (118)	16.1 ± 0.3 (151)
Females	21.9 ± 0.3 (200)	17.2 ± 0.3 (155)

Age in months ± SEM. Animal numbers in parentheses.
Differences between GF and CV and between male and female mice are statistically significant with p values ≤0.01. Data calculated from Gordon et al.[2]

Not only did the GF S-W mice obviously outlive their CV counterparts in this study, but the GF males appeared to outlive the GF females. The CV mice, on the other hand, show the usual trend: females outliving males. A small study conducted at approximately the same time with a limited number of L-W rats had already shown a similar trend (Wostman, B. S., unpublished data), which was confirmed later by the actuarial data drawn from the L-W GF and CV rat colonies.[21]

Pathology in the GF S-W mice at the time of death was quite different from that seen in the CV mice. Whereas in the CV mice respiratory infection was found to be the major cause of death (38%), in the GF mice intestinal atonia ("...resulting in an apparent cessation of gut movement") with 36%, and intestinal volvulus with 9%, were listed as the main causes. Kidney lesions, presumably caused by the high casein content of the L-462 diet,[20] occurred in about 13% of both groups.[2]

It should be mentioned that Walburg and Cosgrove[22] found a marginally but not significantly longer life span in GF ICR mice. However, this study was conducted with far fewer animals than the Lobund study.

In the recent Lobund Aging Study similar but not quite as pronounced differences in life span were seen between GF and CV male L-W rats[23] (Figure 2). In this study the natural ingredient diet L-485 was fed, which does not contain casein[24] (for details see Chapter V) and causes few or no kidney lesions.[23] Also, to insure improved comparability, both

GF and CV rats were housed in isolators, although for the CV rats the isolation was not complete. "Conventional" isolators were opened regularly for introducing food and water and to clean cages. At those times the CV rats were weighed outside the isolator. These management procedures increased the median life span of male CV L-W rats by 2 to 3 months over earlier values of around 24 months observed in our and other[25] laboratories.

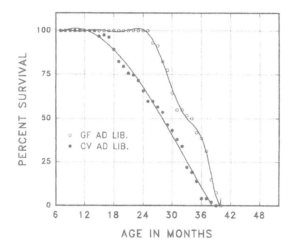

FIGURE 2
Percent survival of male germfree and conventional Lobund-Wistar rats maintained on diet L-485.[24]

To what extent the increase in life span of GF rats and GF mice is related to their lower oxygen consumption,[26,27] and to their, allegedly related, lower heart and liver weights,[15,28] remains a matter of speculation. However, in a recent, unrelated study of vitamin B_6 metabolism, it was established that young, growing GF L-W rats fed a casein-starch diet also showed a significantly lower *ad libitum* food intake than their CV counterparts, although they still grew at a comparable rate.[29] This phenomenon has now been confirmed in male GF L-W rats maintained on natural ingredient diet L-485 (Wostmann, B. S., unpublished data) and had been noticed earlier in GF mice.[30] Together, these data suggest a more efficient use of dietary energy by the young GF rat and GF mouse. This in turn could relate to the catecholamine-inhibitory substances produced in the enlarged cecum of the GF rodents,[31,32] which presumably reduce the energy-dissipating action of brown adipose tissue (see Chapters III and IV). The resulting smaller hearts and livers of these GF animals, once established, appear to remain smaller than those in the CV animals well into old age, a phenomenon which could explain the aforementioned lower adult body weights of the *ad libitum*-fed GF rats.

This impression is reinforced by the fact that when dietary intake was restricted to 12 g L-485 diet per day, or approximately 70% of *ad libitum* intake, the body weight of these GF rats now slightly exceeded that of their CV counterparts during their entire life span. Even when body weight was corrected for cecal enlargement, in the adult animals this small difference remained significant (Table 2). Thus, when limited intake enforced maximal dietary efficiency, GF rats were able to attain slightly higher body weights than their CV counterparts.[33]

TABLE 2
Body Weight in the Second Year of Life, and Survival Percentage at the End of That Period, of Male Lobund-Wistar Rats Fed *ad libitum* or Restricted to 12 g diet L-485[24] per day

		Body weight[a]	Survival (%)
Germfree	Ad lib.	425 ± 11[b]	100
	12 g/d	303 ± 4[c]	100
Conventional	Ad lib.	466 ± 12	69
	12 g/d	285 ± 5	95

[a] Body weights corrected for weight of cecal contents; data are averages ± SEM.
[b] Significantly lower than comparable conventional value; $p = 0.02$.
[c] Significantly higher than comparable conventional value; $p = 0.01$.

The Lobund Aging Study provides challenging data on the effects of dietary restriction on longevity. Whereas the *ad libitum*-fed GF rats show a definite increase in median life span over their CV counterparts, but little extension of maximal attainable age, the diet-restricted GF rats not only show a further increase in median life span, but also a substantial increase in the maximal attainable age. However, the diet-restricted CV rats now show a survival curve almost identical to that of the diet-restricted GF rats, especially in the case of the older animals (Figure 3). The data suggest that dietary restriction improves the functional conditions of the restricted animals to the extent that the challenge of a potentially life-shortening microflora is easily met. Similar results have been reported by Sasaki and co-workers, who restricted GF ICR mice to 80% of *ad libitum* dietary intake.[34]

CONCLUSIONS

Thus, in general it can be stated that *ad libitum*-fed GF rats and GF mice grow as well as comparable CV rats and mice, although maximum

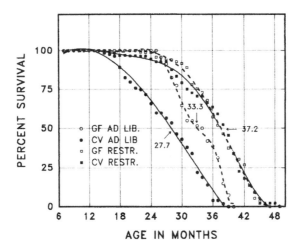

FIGURE 3

Percent survival of *ad libitum*-fed and diet-restricted male germfree and conventional Lobund-Wistar rats maintained on diet L-485;[24] 50% survival age indicated.

body weight may be slightly less. Reproduction may be negatively affected not only by cecal enlargement, but also by the absence of a steroid-modifying microflora. Cecal enlargement conceivably could also affect life span in a negative way. This potential effect notwithstanding, *ad libitum*-fed GF mice and GF rats appear to outlive their CV counterparts.

However, Sasaki and co-workers recently reported that association of GF ICR mice with a controlled and apparently beneficial microflora could extend the life span of both *ad libitum*-fed and diet-restricted mice. These gnotobiotes had been associated with a flora consisting of *Escherichia coli, Klebsiella oxytoca, Proteus mirabilis, Staphylococcus aureus, S. xylosis, Bacillus longum, E. rectale, Fusobacterium prausnitzii, Lactobacillus acidophilus, L. plantarum,* and *B. distasonis.* The presence of this microflora increased the life span of both groups by about 20%, from 636 to 760 days in the *ad libitum*-fed group and from 759 to 892 days in the diet-restricted animals.[35] Apparently, no negative effects of the presence of this microflora were observed. This would be a first indication that the presence of a well-adjusted microflora per se can extend the life span of an animal.

As far as GF gerbils, GF hamsters, and other GF animals are concerned, not enough data are available to make any definite statement. A limited number of GF gerbils, those not lost to intestinal strangulation caused by their greatly enlarged ceca, have lived to an average age of 13 months in apparent good health.[9] As far as GF rats and GF mice are concerned, where we have a plethora of data, it can be stated that growth, reproduction, and life span are all well within acceptable ranges.

REFERENCES

1. Iwasaki, K., Gleiser, C. A., Masoro E. J., McMahan, C. A., Seo, E., and Yu, B. P., The influence of dietary protein source on longevity and age-related disease processes of Fischer rats, *J. Gerontol.*, 43, B5, 1988.

2. Gordon, H. A., Bruckner-Kardoss, E., and Wostmann, B. S., Aging in germfree mice: life tables and lesions observed at natural death, *J. Gerontol.*, 21, 380, 1966.

3. Snyder, D. L. and Wostmann, B. S., Growth rate of male germfree Wistar rats fed *ad libitum* or restricted natural ingredient diet, *Lab. Anim. Sci.*, 37, 320, 1987.

4. Henderson, J. D. and Titus, J. L., Growth rate and morphology of germfree and conventional mice, *Mayo Clin. Proc.*, 43, 517, 1968.

5. Heneghan, J. B., Longoria, S. G., and Cohn, I., The growth of germfree beagles, *Am. Surg.*, 34, 82, 1968.

6. Küster, E., Die Gewinnung, Haltung und Aufzucht keimfreier Tiere and ihre Bedeutung für die Erforschung natürlicher Lebensvorgänge, *Arb. Kais. Gesundh. Amtes*, 48, 1, 1915.

7. Coates, M. E., Fuller, R., Harrison, G. F., Lev, M., and Suffolk, S. F., A comparison of the growth of chicks in the Gustafsson germfree apparatus and in a conventional environment, with and without dietary supplements of penicillin, *Br. J. Nutr.*, 17, 141, 1963.

8. Gordon, H. A., Is the germfree animal normal? A review of its anomalies in young and old age, in *The Germfree Animal in Research*, Coates, M. E., Ed., Academic Press, London, 1968, chap. 6.

9. Wostmann, B. S., Beaver, M., Bartizal, K., and Madsen, D., Gnotobiotic Gerbils, in Proc. 4th Int. Symp. Contamination Control, Washington, D.C., Vol. 4, 1978, 132.

10. Saito, M. and Nomura, T., Production of germfree animals. I. Small mammals, in: *The Germfree Animal in Biomedical Research*, Coates, M. E. and Gustafsson, B. E., Eds., Laboratory Animals Ltd., London, 1984, chap. 2.

11. Umehara, K., Tazume, S., Hashimoto, K., and Sasaki, S., Increased reproduction of germfree mice contaminated by bacteria, in *Experimental and Clinical Gnotobiology, 10th International Symposium on Gnotobiology*, Heidt, P. J., Vossen, J. M., and Rusch, V. C., Eds., *Microecol. Ther.*, 20, 441, 1990.

12. Gustafsson, B. E., Gustafsson, J. A., and Sjövall, J., Intestinal and cecal steroids in germfree and conventional rats, *Acta Chem. Scand.*, 20, 1827, 1966.

13. Gustafsson, B. E., Gustafsson, J. A., and Sjövall, J., Steroids in germfree and conventional rats, *Eur. J. Biochem.*, 4, 568 and 574, 1968.

14. Eriksson, H., Gustafsson, J. A., and Sjövall, J., Steroids in germfree and conventional rats, *Eur. J. Biochem.*, 6, 219, 1968.

15. Gordon, H. A., Bruckner-Kardoss, E., Staley, T. E., Wagner, M., and Wostmann, B. S., Characteristics of the germfree rat, *Acta Anat.*, 64, 301, 1966.

16. Reddy, B. S., Pleasants, J. R., Zimmerman, D. R., and Wostmann, B. S., Iron and copper utilization in rabbits as affected by diet and germfree status, *J. Nutr.*, 87, 189, 1965.

17. Pleasants, J. R., Reddy, B. S., Zimmerman, D. R., Bruckner-Kardoss, E., and Wostmann, B. S., Growth, reproduction and morphology of naturally born, normally suckled germfree guinea pigs, *Z. Versuchstierk.*, 9, 195, 1967.

18. Young, W. C., Goy, R. W., and Phoenix, C. H., Hormones and sexual behavior, *Science*, 143, 212, 1964.

19. Pollard, M., Senescence in germfree rats, *Gerontologia*, 17, 333, 1971.

20. Wostmann, B. S., Nutrition of the germfree mammal, *Ann. N.Y. Acad. Sci.*, 78, 175, 1958.

21. Pollard, M. and Wostmann, B. S., Aging in germfree rats: The relationship to the environment, diseases of endogenous origin, and to dietary modification, in: *8th Symposium of the International Council for Laboratory Animal Science (ICLAS), Vancouver, 1983*, Archibald, J., Ditchfield, J., and Rowsell, H. C., Eds., Gustav Fischer, New York, 1985, 181.

22. Walburg, H. E. and Cosgrove, G. E., Aging in irradiated and unirradiated germfree ICR mice, *Exp. Gerontol.*, 2, 143, 1967.

23. Snyder, D. L., Pollard, M., Wostmann, B. S., and Luckert, P., Life span, morphology, and pathology of diet-restricted germfree and conventional Lobund-Wistar rats, *J. Gerontol.*, 45, B52, 1990.

24. Kellogg, T. F. and Wostmann, B. S., Stock diet for colony production of germfree rats and mice, *Lab. Anim. Care*, 19, 812, 1969.

25. Hoffman, H. J., Survival for selected laboratory rat strains and stocks, in Development of the Rodent as a Model System of Aging, Gibson, D. C. *et al.*, Eds., DHEW Publ., U.S. Government Printing Office, Washington, D.C., 1978, 19.

26. Desplaces, A., Zagury, D., and Sacquet, E., Étude de la fonction thyroidienne du rat privé de bactéries, *C. R. Acad. Sci.*, 257, 756, 1963.

27. Wostmann, B. S., Bruckner-Kardoss, E., and Knight, P. L., Cecal enlargement, cardiac output, and O_2 consumption in germfree rats, *Proc. Soc. Exp. Biol. Med.*, 128, 137, 1968.

28. Wostmann, B. S., Bruckner-Kardoss, E., and Pleasants, J. R., Oxygen consumption and thyroid hormones in germfree mice fed glucose-amino acid liquid diet, *J. Nutr.*, 112, 552, 1982.

29. Wostmann, B. S., Snyder, D. L., and Pollard, M., The diet-restricted rat as a model in aging research, *Microecol. Ther.*, 17, 31, 1987.

30. Yamanaka, M., Iwai, H., Saito, M., Yamauchi, C., and Nomura, T., Influence of intestinal microbes on digestion and absorption of nutrients in diet and nitrogen retention in germfree, gnotobiotic and conventional mice. I. Protein and fat digestion and nitrogen retention in germfree and conventional mice, *Jpn. J. Zootech. Sci.*, 43, 272, 1972.

31. Baez, S. and Gordon, H. A., Tone and reactivity of vascular smooth muscle in the germfree rat mesentery, *J. Exp. Med.*, 134, 846, 1971.

32. Wostmann, B. S., Reddy, B. S., Bruckner-Kardoss, E., Gordon, H. A., and Singh, B., Causes and possible consequences of cecal enlargement in germfree rats, in *Germfree Research. Biological Effects of Gnotobiotic Environments. Proceedings of the IV International Symposium on Germfree Research*, Heneghan, J. B., Ed., Academic Press, New York, 1973, 261.

33. Wostmann, B. S., Snyder, D. L., and Johnson, M. H., Metabolic effects of the germfree state in adult, diet-restricted male rats, in *Experimental and Clinical Gnotobiology. Proceedings of the Xth International symposium on Gnotobiology*, Heidt, P. J., Vossen, J. M., and Rusch, V. C., Eds., *Microecol. Ther.*, 20, 383, 1990.

34. Tazume, S., Umehara, K., Matsuzawa, H., Hashimoto, K., and Sasaki, S., The effect of dietary restriction on germfree mice, *Microecol. Ther.*, 20, 373, 1990.

35. Tazume, S., Umehara, K., Matsuzawa, H., Hashimoto, K., and Sasaki, S., The Effects of Microbial Status and Food Restriction on Longevity of Mice, Paper presented at the XIth Int. Symp. Gnotobiology, Belo Horizonte, Brasil, 1993 (Abstr.).

Chapter III

ANATOMY, MORPHOLOGY, AND FUNCTION OF THE GASTROINTESTINAL SYSTEM

GENERAL ASPECTS

The majority of gastrointestinal data have been obtained from GF rats, while reliable data are also available from GF mice. Beyond this, a smaller body of data is available on GF rabbits,[1,2] guinea pigs,[3,4] pigs,[5,6] and chickens.[7,8] This chapter will emphasize GF rats and GF mice.

All data agree that the major effects of the total absence of a bacterial microflora are expressed mainly in two organ systems:

1. The gastrointestinal system, where normally an intense interaction takes place between the host and its microflora, especially in the more distal parts of the intestinal tract. The microflora represents an entity with a weight comparable to one of the larger organs of the body.[9] According to Ducluzeau, the digestive tract of a mammal contains a number of bacteria 10 times greater than the actual number of cells forming that mammal,[10] and comprises at least 28 different genera.[11]

2. The immune system, in part in close proximity to the alimentary tract, but mainly monitoring the blood and lymph. It comprises not only immune defenses, but may also influence via its secretagogues a wide range of other functions, including some that may then affect the gastrointestinal tract.[12-15]

Some influence has been reported on the skin of the GF animal, although visual observation of the healthy GF animal does not detect this. Gordon found a lower value for the regional blood flow in the skin,[16] but all major adaptations to the GF environment which have become obvious appear to be related to functional changes in the alimentary and immune systems.

In general, the mouth and esophagus show little effect of the absence of bacteria (except for the much smaller submandibular lymph nodes), but no true dental caries has ever been found in a GF animal.[17,18] Even in the stomach little change is seen, although slower emptying has been reported.[19,20] The small and large intestine, but especially the cecum, demonstrate the absence of the microflora. The differences in the small and large intestine largely can be explained by the absence of the "stimulatory" action of the locally residing microflora elements, presumably via products of bacterial metabolism. Peristalsis was found to be considerably slower in the absence of a microflora. Abram and Bishop found lower propulsion rates in GF mice,[19] whereas Sacquet et al. found similar differences in rats.[20] In the GF rat differences were found to depend, at least in part, on the composition of the diet.[21] Pen and Welling point to a plausible cause of the slower propulsion rate, since they found cholecystokinin levels in the gut of GF mice lower than in their CV counterpart.[22] A recent study by Husebye et al. again confirms the differences in gut mobility, but stresses the role of the microflora in stimulating the migrating motor complex via its biochemical effecters.[23] Caenepeel et al.[24] studied the interdigestive myoelectric complex (EDMEC), a cyclic electromyographic pattern which migrates down the small and the large bowel. They found migration times of 20 and 102 min for GF, but only 15 and 75 min for CV rats, again indicating the influence of the gut microflora on intestinal physiology. This difference in peristalsis could contribute to the greater absorptive capacity demonstrated for a number of materials, including calcium. The significantly greater retention of Ca by the GF rat in turn would explain its somewhat heavier and denser skeleton.[25]

In the cecum the absence of microbial digestion of the intestinal mucus leads to an accumulation of this material, which proved to be the major cause, directly and indirectly, of the cecal enlargement in GF rodents, rabbits, and guinea pigs. Conditions in the GF cecum also promote the occurrence of materials which, once transmitted to the bloodstream, appear to greatly affect the function and metabolism of the animal. Morphologically, this appears to be expressed in the smaller hearts and lungs of the GF rodent,[26-28] a reflection of its lower resting O_2 use,[29-31] apparently caused by the young GF rodent's more efficient energy metabolism (see Chapter IV and Chapter V).

STOMACH

As mentioned earlier, no differences in morphology or function of the stomach have been found in the GF state, except for a somewhat slower emptying rate.[19] This, however, may be diet dependent.[20] Histologically, no differences have been observed between the gastric mucosae of GF and CV mice.[32]

SMALL INTESTINE

It soon became apparent that whereas the length of the GF rat small intestine was comparable to that of its CV equivalent, its weight was considerably less. This weight differential is quite diet dependent: about 15% on the early casein-starch diets,[26] but more than 30% on the natural ingredient diet L-485* (for diet compositions see Chapter V). The difference occurs mainly in the lamina propria, which in the GF animal is less well developed than in its CV counterpart. Only a few lymphocytes and histiocytes are found; leukocytes are scarce compared to the lamina propria of the healthy CV animal, whose gut can be considered to be always in a state of what Gordon[33] called "physiological inflammation". In GF animals Peyer's patches are small, with few reactive centers, low mitotic activity, and a rather small number of plasmacytes.[34-36]

The description of **villus morphology** presents a somewhat confusing picture since early data appear to be influenced to a considerable extent by diet, and presumably by the intestinal microflora of the CV animals, which may have differed extensively from laboratory to laboratory. While Gordon and Bruckner-Kardoss[37] found villi to be smaller in the GF than in the CV rat, Galjaard et al.[38] found them to be longer. Komai and Kimura[39] added graded levels of cellulose to the diet of GF and CV ICR/JCL mice and observed that while on the fiberless diet villi in the mid-jejunum of the GF mice were approximately 35% longer than in the CV mice; fiber addition caused the GF villi to become shorter, whereas the villi of the CV mice became longer. At a fiber content of 30% the difference in villus length had disappeared. In 1977 Abrams summarized the available data by stating that, in general, villi of the distal small intestine have been found to be smaller in the GF rat and the GF mouse, while in the GF duodenum they are longer.[40] There is a consensus, however, about the ratio of villi to crypt enterocytes, it being higher in GF than in CV animals, the difference becoming smaller in the more distal parts of the small intestine.[41] Recently Uribe et al. reported moderate hyperplasia of the villi in the proximal small intestine of the GF rat in the absence of any changes of the crypt population.[42] They found the cell birth rate and proliferative profile similar in GF and CV animals, comparing well with the latest literature data.[43] Microvilli appear to be somewhat longer in the GF state,[44] but as demonstrated in the GF mouse, they lack the fucolipids that are found in the CV mouse brush border.[45]

Gordon and Bruckner-Kardoss,[37] Meslin et al.,[46] and also Stevens[47] found the total surface area of the GF rat small intestine to be considerably smaller than in the CV rat. This effect was most pronounced in the mid and lower parts of the small intestine. The difference found by Meslin

---

* Data taken from the Lobund Aging Study.

et al. in Fischer rats (14%) is considerably less than Gordon and Bruckner-Kardoss reported for the Lobund-Wistar (L-W) rat (30%), but combining all available rat data it would appear that the total mucosal surface of the GF rat approximates 75% of that of the CV animal.[41]

Studies with GF and GN piglets produced results largely comparable to the above. The GF piglets showed long, slender, well-developed villi, in contrast to the smaller, somewhat irregularly shaped villi of their CV counterparts.[48] Miniats and Valli[49] report the weight of the GF small intestine to be about 50% of that of the CV piglets. The GF animals showed reduced thickness and cellularity of the lamina propria and muscularis with poorer development of the associated lymph tissue. However, Heneghan *et al.*[50] report that the mucosal surface area of the small intestine of the GF piglet is comparable to or possibly slightly larger than that in CV piglets.

The **immune elements** in the gut occur organized in the Peyer's patches, and are diffusely distributed in the lamina propria. Although Peyer's patches can be found throughout the small and large intestine of the GF rat and the GF mouse, they are small, and less effective in the uptake of latex particles.[51] As mentioned earlier, fewer immune elements are found in the lamina propria. In the GF rat, Knight and Wostmann[52] found about 20% of the amount found in the CV animal.[34] Similar results have been reported for GF mice.[53] Crabbé *et al.* later gave a more detailed description of the immunohistochemical picture of the gut of the GF mouse, essentially confirming the earlier observations.[54]

Paneth cells, supposedly involved in the regulation of the intraluminal microflora within the intestinal crypt region, appear to be smaller but more numerous in its absence[55] (see also Chapter VII). Satoh and Vollrath[56] describe small variations in the ultrastructure of Paneth cells of GF rats, and stress the release of their secretory granules upon the introduction of a CV microflora. They find no evidence that these cells are actively involved in the phagocytosis of bacteria. Later studies by this group stress the fact that, besides lysozymes, Paneth cells appear to regulate the bacterial milieu of the intestine by releasing secretory granules containing IgA into the crypt lumen.[57]

Both microflora and dietary components appear to affect the **renewal rate of the intestinal mucosa**. To an extent, this could involve polyamines produced by the microflora or originating from the diet.[58,59] In the absence of a microflora, the transit time of the enterocytes from the crypts of Lieberkuhn to the extrusion zones at the tips of the villi may be increased by as much as a factor 2.[60-63] Lesher *et al.*[64] found the size of the proliferative pool of GF mice to be reduced by 37%, and the generation time of the proliferative cells by 25%, resulting in a decrease in cell delivery of almost 50%. Combining the data of Heneghan's review,[41] the ratio GF/CV for the villus length traveled in the mouse ilium is 37/84 or 0.44. In other words, it takes the enterocyte in the GF mouse about twice as long to reach the top of the villus as it does in the CV mouse (enterocyte transition time). Observations largely similar to the above, but not as pronounced as in the GF mouse, have been reported by Rolls et al.[65] for the GF chicken. At the most recent meeting of the International Symposium on Gnotobiology, Alam et al.[66,67] reported similar but somewhat more detailed data on enterocyte kinetics of GF and CV AGUS rats. Their data show less of a difference between the GF and CV ileum, but a more outspoken dissimilarity in the colon. They also point to the potential role of glucagon in determining enterocyte kinetics.[68]

As a result, the average enterocyte of the germfree animal is older than the corresponding cell of the CV animal. Since age plays a role in the content of a number of hydrolytic enzymes of the enterocyte,[69] this could explain the much higher levels of disaccharidases,[2,70-73] ATP-ase,[74] and alkaline phosphatase[72,73] found in the intestinal mucosa of the GF mouse, rat, and rabbit. However, Komai and Kimura were able to substantially decrease enterocyte transit time by feeding increasing amounts of cellulose in the diet of GF ICR/JCL mice for 24 days and did not find a corresponding decrease in alkaline phosphatase.[39] Also, Miyazawa and Yoshida were not able to detect any difference in either alkaline phosphatase (EC 3.1.3.1) or phytase (EC 3.1.3.8), supposedly very similar or possibly identical enzymes, in the upper small intestinal mucosa of GF and CV Fischer rats.[75] Also, the above differences appear not to exist between GF and CV chickens.[76]

When GF rodents are exposed to a "conventional" microflora, the aforementioned differences would disappear in about 1 week,[50] although Olson and Korsmo[71] state that in their studies with GF Sprague-Dawly rats elevated sucrase-isomaltase levels persisted for at least 2 weeks after housing the animals with CV rats. This may just indicate that "conventionalizing" via contact with adult CV animals for a few weeks may not always lead to the establishment of a microflora comparable in all details to the microflora that gradually builds itself in the newborn and young CV animal.

During the first international meeting of scientists involved in the study of germfree life, organized by the New York Academy of Sciences in 1958, Phillips and Wolfe reported that in the greatly enlarged cecum of the GF guinea pig the **oxidation-reduction potential** measured in its contents was significantly more positive than in the contents of the CV guinea pig.[77] Later studies by Wostmann and Bruckner-Kardoss found this difference in the rat to be +250 to +300 mV. When GF rats were inoculated into the cecum with the cecal contents of CV rats, 4 h later oxidation-reduction values were comparable to those in CV rats, indicating that the more positive values in the GF animals were directly related to the absence of an intestinal microflora.[78,79] Although the author is not aware of any measurements in the small intestine, we may presume that similar, although possibly smaller differences in the oxidation-reduction potential will exist in the more distal parts. Along the same lines we could speculate that the observation by Bornside et al., who determined O_2 and

CO_2 tension in the colon and found O_2 tension somewhat lower in CV than in GF rats, while CO_2 tensions were higher and dependent on the microflora composition, might also hold for the distal small intestine.[80] The pH of intestinal and cecal contents was found to be somewhat more to the acid side in animals housing an intestinal microflora.[78,81]

Digestion and absorption of ingested materials are the obvious functions of the intestinal tract. These processes are affected by intestinal propulsion, the intestinal secreta, and the effects of the intestinal microflora thereupon. In the small intestine, bile, pancreatic enzymes, and secreta from the intestinal wall all are involved in digestion and absorption. The availability of GF animals has made it possible to evaluate the effects of the microflora on these functions. The data reveal the extent to which the intestinal enzymes are broken down by the microflora. Also, the intestinal bile acid composition is strongly affected, both qualitatively and quantitatively.

The effect of the GF state on pancreatic enzymes was first studied in the chicken. Lepkovsky *et al.*[82] found higher levels in the lower gut of GF than of CV chickens. Later studies established that whereas the microflora apparently does not affect the concentration and secretion of pancreatic trypsin, chymotrypsin, amylase, or elastase, these enzymes undergo a progressive microbial inactivation upon entering the small intestine and subsequently passing through the entire intestinal tract (Table 1). As bacterial counts increase from the small to the large intestine in the CV animal, host intestinal enzyme activity decreases accordingly.[83,84] Only the pancreatic lipase appears to be more resistant to microbial attack.[83] In GF animals, on the other hand, enzyme activities remain relatively high, and may therefore result in more complete substrate degradation. This would explain the increased levels of free amino acids and free fatty acids found in the fecal and cecal content of the GF rat.[85]

TABLE 1
Effect of the Intestinal Microflora on Pancreatic Enzymes.
Conventional Rat Values Expressed as Percentages
of Comparable Values in Germfree Rats

	Lipase	Amylase	Trypsine	Chymotrypsin
Pancreas	97	91	101	90
Small intestine	100	43	60	57
Cecum	32	59	36	38
Large intestine	10	9	29	34

Data calculated from Reddy *et al.*[83]

In the GF rat and the GF mouse the composition of the **intestinal bile acids** reflect that of the biliary secretion. In the small intestine their total concentration is several times higher than in the CV animal[86-88] and consist almost exclusively of taurine-conjugated primary bile acids (Table 2).[89] Thus, in the GF L-W rat small intestine bile acids largely consist of about equal parts of taurine-conjugated cholic and β-muricholic acid, while no ω-muricholic acid is found.[90] In the CV animal, on the other hand, around 70% of the intestinal bile acid pool consists of cholic acid, the remainder is 20% β-muricholic acid, 5% hyodeoxycholic acid, and a few percent of deoxycholic and ω-muricholic acids. During their passage through the CV lower small intestine these bile acids are gradually deconjugated and otherwise modified, until in the cecum they are almost totally in the deconjugated form.

TABLE 2
Bile Acids in the Contents of the Third Quarter
of the Small Intestine of Male Germfree
and Conventional Lobund-Wistar Rats

	Percent of total	
	Germfree	Conventional
Lithocholic	—	0.9 ± 0.4
Deoxycholic	—	2.5 ± 0.4
Chenodeoxycholic	0.8 ± 0.1	0.5 ± 0.2
Cholic	51.1 ± 2.1	69.8 ± 3.2
Hyodeoxycholic	—	5.4 ± 0.8
β-Muricholic	47.9 ± 2.1	18.8 ± 1.9
ω-Muricholic	—	1.8 ± 0.3
Total (in mg)	22.2 ± 2.8	5.1 ± 1.6

Data are averages ± SEM, taken from Madsen *et al.*[90]

Riottot *et al.*[91] consider the slow propulsion rate in the GF rat to be the main factor that causes its larger bile acid pool, whereas Gustafsson and Norman demonstrated that, especially in the GF rat, diet has a significant effect on the eventual elimination of bile acids from the body.[92] The potential effect of the enterohepatic bile acid pool on intestinal function was illustrated by Ranken *et al.*,[93] who found that upon adding free cholic acid to the diet of GF mice the turnover of enterocytes would approach CV values. This suggested that free cholic acid might be a main factor in determining the enterocyte renewal rate. Later studies by Mastromarino and Wilson appear to confirm that the relatively large amount of deconjugated cholic acid present in the gut of the CV mouse could be the reason for its higher mucosal turnover rate.[94] All this indicates the substantial influence microflora and diet have, via the intestinal bile acids, on intestinal physiology. For a further discussion of the effects of the qualitatively and quantitatively different bile acid pool of the GF animal see Chapter IV.

Intestinal mucus produced by the goblet cells forms a protective layer over the intestinal mucosa and becomes mixed with the intestinal contents. Data from rats, dogs, and piglets indicate that in the GF state these animals show 35 to 50% more goblet cells than their CV counterparts.[41,49,57] Umesaki *et al.* established that, as a result, the GF mouse produced more mucin, with an apparently higher hexose content, the latter observation presumably related to the higher fucosyltransferase activity found in microvillus membrane preparations.[95] In the GF rodent this material eventually accumulates in the cecum, where its water-holding capacity can be seen as the main reason for cecal enlargement (see below). Gordon and Wostmann,[96] and later Gustafsson and Carlstedt-Duke,[97] have demonstrated that in the CV animal these mucins are extensively degraded by the intestinal microflora.

Absorption from the small intestinal lumen in essence is determined by three factors: the functional capabilities of the intestinal musosa as a whole and of the individual enterocyte, the conditions in the lumen of the gut, and to an as yet undetermined extent by the water-holding mucus layer covering the microvilli. Meslin and Sacquet[98] found the microvilli of the GF rat to be somewhat smaller than in the CV animal. This may have reduced the size of the total "unstirred" water layer to some extent, possibly making absorption of nutrients somewhat easier. As far as enterocyte function itself is concerned, most data appear to point to a greater, or at least comparable, passive or carrier-facilitated potential of the GF gut.[99] Easier paracellular absorption could also be a factor here, but the data may also be affected by the slower intestinal propulsion of the GF animal, which in some cases may enhance absorption. On the other hand, no difference seems to exist between active ATP-driven absorption of materials like Na^+, K^+, or glucose.[100,101] However, as soon as the rate of ATP-driven absorption is not rate-determining, conditions in the lumen and the rate of peristalsis could play a significant and possibly overriding role. Although the somewhat larger number of goblet cells found in the GF intestine (see above) suggests increased production of intestinal mucins, to our knowledge no mention has been made of any effect of a possible change in the unstirred layer on intestinal absorption.

Earlier, we mentioned the differences in pH, oxidation-reduction potential, O_2 and CO_2 tension, bile acid composition and concentration, and the concentration of intestinal enzymes. All these may affect the condition in which the various materials in the intestinal tract are available for the mechanics of absorption.

Recently, it has become obvious that immune cell secretagogues may have a significant effect on water and mineral absorption.[13] Cells involved here are the phagocytes, mast cells, and possibly T lymphocytes, all of which are much more abundant in the intestinal tissue of CV animals than in GF animals. This may be the reason that the histamine content in the wall of the small intestine of GF rats and GF mice was found to be lower than in their CV counterparts.[102] On the other hand, 5-hydroxytryptamine levels were higher in the GF intestinal tract,[102,103] presumably because of a higher absorption rate of its precursor tryptophane.[104] In a recent paper Nogueira and Barbosa[105] described their immunocytochemical study of the intestinal endocrine cells of GF and CV CFW mice. They found evidence of highly active 5-hydroxytryptamine- and neurotensin-producing cells in GF mice, besides active enteroglucagon- and peptideYY-positive cells. No difference in somatostatin-positive cells was found, but surprisingly, no cholecystokinin- or neurotensin-positive cells were seen.

It is not clear to what extent the absence of microbial production of histamine or the lower number of mast cells account for the lower GF levels of this material with its potential to influence Cl^- and water transport.[13] In the case of 5-hydroxytryptamine it would seem that local production after enhanced absorption of tryptophane overrides the lower mast cell content. It is obvious, however, that both substances may have an influence on water and mineral absorption. By the same token it is likely that alterations in absorption may be brought about by the many factors secreted by microflora- or diet-stimulated lymphocytes.

The aforementioned abundance of pancreatic amylase and proteolytic enzymes in the GF small intestine in the absence of a degrading microflora assures that dietary carbohydrate and protein utilization will at least be as good or possibly better than in the CV animal (Table 1). Although Reddy et al.[83] found no difference in small intestinal lipase, the much higher (conjugated) bile acid content of the GF rat[86] and the GF mouse[87,106] could be a factor in the somewhat higher utilization of dietary fat by the GF rodent.[107]

Uptake of disaccharides by the enterocytes will be determined by the disaccharidases in the brush border, and the subsequent ATP-driven transport of resulting monosaccharides across the cell. Since disaccharidase activity of the "older" enterocyte of the GF animal is greater than that of its CV counterpart[70-72] this makes active transport of the resulting monosaccharides, for which no difference has been found, the controlling factor. As mentioned, no difference was found in the uptake of glucose. Xylose, on the other hand, which is taken up by passive and carrier-facilitated absorption, was removed from the GF small intestine at approximately twice the rate found in the CV animal.[100] This was later confirmed by the observation that urinary excretion of gavaged xylose was much higher in GF than in CV rats.[108]

Mineral absorption presents a much more varied and sometimes confusing picture. Absorption of many minerals may, directly or indirectly, be influenced by the intestinal propulsion rate, intestinal pH, oxidation-reduction potential, bile acids, and complexing materials like phytic acid. There appears to be a consensus that calcium and magnesium are absorbed much more efficiently in GF rats and mice than in their CV counterparts, and that this may lead to the somewhat more dense and heavier bone structure of the GF animal.[25,99] Data reported by Reddy et al. strongly suggest a more efficient uptake in the small intestine.[25] They relate this more efficient uptake to the increased levels of intestinal brush border Ca^{2+}- and Mg^{2+}-stimulated ATPases, together with the higher levels of alkaline phosphatase in the GF rat. This they ascribe to the aforementioned longer life span of the GF enterocyte.[109] On the other hand, Andrieux and Sacquet, feeding ^{45}Ca with the diet, together with TiO_2 as an unabsorbable marker, found evidence of increased Ca absorption in the lower gut of the GF rat.[110] In either case, the higher intestinal bile acid concentrations of the GF rat would promote micelle formation with mineral-carrying lipid complexes, thereby enhancing absorption. This appears to be in agreement with the observation of Demarne et al. that the fecal excretion of Ca soaps is lower in GF than in CV rats.[107] Easier paracellular absorption through the thinner lamina of the GF animal could be another factor in the increased absorption of Ca and Mg. Although P absorption was not affected, the increased Ca and Mg absorption appeared to lead, via increased bone formation, to a greater retention of P to maintain Ca homeostasis, as reflected in the lower urinary P excretion of the GF animal.[25]

No significant difference was found in the absorption and retention of Zn by GF and CV L-W rats, although the data for the GF animals tended to be somewhat higher.[111] A recent study indicates that Zn uptake may be largely a facilitated absorption via a protein called cysteine-rich intestinal protein (CRIP).[112] Earlier, Smith et al. had obtained results suggesting that GF rats require less Zn than their CV counterparts,[113] as has been demonstrated for a number of other nutrients.[114] Combined, these facts may explain the approximately 20% higher serum Zn levels found in the GF rats.[115]

Though the various factors that control Fe absorption have not been totally elucidated, it seems obvious that the GF state with its differences in pH, oxidation-reduction potential, and potentially complexing materials, may affect Fe absorption in a negative way. Feeding GF rabbits a diet with an ample supply of mineral Fe resulted in severe anemia and the eventual death of some of the animals, whereas their CV counterparts thrived on the same diet. Either conventionalization by exposure of the GF rabbits to a "normal" microflora producing a more negative intestinal oxidation-reduction potential, or replacing part of the inorganic Fe supplement by adding a source of organically bound Fe such as soybean meal to replace part of the diet's cereal content, alleviated all symptoms within a period of 4 weeks.[1] This result was obtained with a diet that now contained even less total Fe than the original mineral supplement would have provided. Thus, in the rabbit the GF state appeared to affect Fe absorption, especially when Fe was originally presented in mineral form, and a high cereal content of the diet provided ample phytate to reduce Fe availability.

In rats the effect of the GF state on Fe nutrition was much less dramatic. However, it should be kept in mind that the rat is relatively insensitive to dietary factors affecting nonheme Fe absorption.[116] Geever et al. reported a higher absorption by the fasted GF rat after gavage of a solution containing $^{59}Fe^{++}$ sulfate and 0.5% ascorbic acid.[117] Uptake from colony type diets, on the other hand, showed more subtle differences, which again might indicate a lower availability of Fe although these diets were fortified with inorganic iron. Plasma iron values tended to be slightly low,[118] but hemoglobin and hematocrit values of GF and CV rats were generally comparable.* However, in at least one study, storage and distribution of Fe suggested a lower rate of Fe and also of Cu metabolism in the GF animals.[117]

Reddy et al. found no differences in Mn values,[118] but Andrieux et al. reported an increased absorption and retention in GF rats.[119]

CECUM

The animal models that have been studied thus far fall into two classes: animals whose cecum is greatly enlarged in the GF state (rat, mouse, gerbil, hamster, guinea pig, rabbit), and those that do not show much enlargement of the comparable organ (dog, pig, sheep, goat, chicken). The latter group, because of the anatomical structure at the junction of the small and large intestine, develops little or no enlargement, but locally may show similar biochemical parameters resulting from the GF state.[7,99,120]

Cecal enlargement in rodents starts early in life. GF rats show no enlargement after the first week, but at 16 days after birth the cecum is already enlarged fourfold.[121] Eventually enlargement may reach a factor of 8, depending on the nature of the diet. In absolute terms, the cecum in both GF and CV rodents reaches its maximum size in 3 to 4 months, together with the rest of the intestinal system. Expressed in percent of body weight, the cecum of the GF rat maintained on natural ingredient diet L-485[122] reaches a maximum of 10 to 12% of body weight shortly after weaning (Figure 1).

By the time the GF L-W rat has reached its full adult weight of around 500 g, its cecum will represent only 3 to 4% of body weight, compared to less than 1% in the CV animal.[123] Whereas throughout the small intestine there is little difference in the usual dry matter percentage of the intestinal contents of around 25%, this value drops to half in the GF cecum, indicating an extensive influx of water.[16,88,121] Consequently, the wall of the GF cecum shows all the signs of the stress of the resulting distension. Individual smooth muscle cells are very elongated and hypertrophied. The tone of cecal muscle strips was found to be approximately two-thirds of that in the CV rat,[16] possibly because of the absence of stimulatory materials originating from the microflora of a CV animal. Jervis and Biggers[124] mention that the cecal mucosa of the GF mouse is generally thinner, with short, irregular villi covering its surface. A detailed description of the ultrastructure of the cecal wall of the GF rat has been given by Gustafsson and Maunsbach.[125]

---

* The Lobund Aging Study, unpublished results.

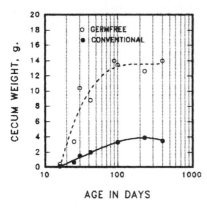 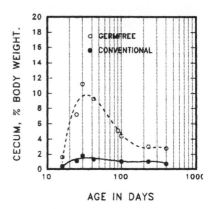

FIGURE 1
Cecal weight, absolute and as a percent of body weight, of germfree and conventional male
Lobund-Wistar rats. Data from the Lobund Aging Study.

The origin of the "enlarged cecum syndrome" can be found to a major
extent in the accumulation of intestinal mucus in the cecum in the absence
of a mucus-degrading microflora. Several factors play a role here. The
accumulating high molecular weight, negatively charged mucopolysac-
charides represent a substantial water-attracting force. Normally this material would be degraded, largely by the microflora. Gordon and Wostmann found considerable amounts of material with a molecular weight
greater than 80 kDa to be present in the supernatant (15,000 × g, 10 min)
of GF cecal contents, whereas hardly any of this material was found in
the CV cecal supernatant.[96] Under CV conditions, the sulfate groups of
this material would, after microbial degradation, be available to take part
in the normal ionic equilibrium between the intestinal contents and the
interior milieu. However, being tied to large macromolecules in the GF
cecum these groups cannot take part, thus driving other negatively
charged ions out of the cecum. Asano was first to show that in GF rat and
GF mouse cecal contents the concentration of Cl^- ions was less than 10%
of that in the cecum of the CV rat and the CV mouse.[126,127] This paucity
of diffusible negative ions apparently makes it impossible for the ATP-
driven Na^+ pump in the GF cecal wall to remove the excess water, even
though the increased potential of this mechanism under GF
conditions[128,129] suggests an adaptation, however insufficient, in an effort
to maintain homeostasis.[88] Asano was able to increase Cl^- availability in
the gut, including the cecum, by feeding GF rats a chloride-yielding resin.
This reduced the size of the cecum significantly, and improved water
absorption from the lower bowel.[130] Gordon and Wostmann demonstrated
that when the cecal contents of the GF rat were replaced *in vivo* with a
saline solution in 10% polyvinylpyrrolidone (PVP; used to measure the
outgoing water flux), normal or even elevated levels of water absorption

from the cecum were observed, together with a substantial reduction in its size.[95] Similar conclusions were reached by Donowitz and Binder.[131]

Another factor which appears to play a role in cecal enlargement is the *in situ* accumulation of urea, which in the cecum of the GF rat is found in approximately the same concentration as in its blood.[132,133] This end product of protein metabolism occurs at approximately the same level in the blood of GF and CV animals (BUN values).[134]* Urea is removed not only by the kidney but also to a major extent via the bile into the intestine, where in CV animals it is almost totally degraded by the microbial ureases in the cecum. In contrast, the GF rat lacks any urease potential.[135] Thus, whereas CV rats hardly excrete any urea with the feces, Harmon et al.[136] report that their GF counterparts may excrete up to 25% of voided urea in that way. The fact that Juhr and Ladeburg[137] could achieve a substantial reduction in cecum size in both GF rats and GF mice by administering urease with the drinking water indicates that urea plays a role in cecal enlargement. Obviously this effect depends on the protein content of the diet.

Thus, while the accumulation of mucins in the GF cecum is a major factor in its enlargement, and urea appears to play an additional role, the GF state also leads to high levels of proteolytic and other enzymes in the absence of a degrading microflora (Table 1). This environment may therefore give rise to metabolic products which might occur under CV conditions but in concentrations that would hardly be of consequence. Combe et al. reported increased levels of hexosamines and free amino acids, reflecting to a large extent a partial enzymatic degradation of the accumulated intestinal mucins.[133,138]

As early as 1965 Gordon reported that the contents of the GF rodent cecum contained materials which affected smooth muscle and which were toxic upon intraperitoneal administration to CV mice.[139] One of these substances was tentatively identified as kallikrein.[140] Somewhat later Baez and Gordon established that the microvasculature of the GF rat is quite refractory to catecholamines, vasopressin, and to some extent angiotensin. Using the microvasculature in the mesentery of the jejunal-ileal region as a test system, they found that approximately 100 times as much norepinephrine was required to produce a standard constriction. Cecectomy of the GF rats at an early age largely abolished the above differences.[141,142] This observation was then related to one of the substances isolated from the supernatant of GF cecal contents — a brown iron-containing material, presumably a degradation product of ferritin originating from desquamated mucosal cells. This material, which was designated α-pigment,[143,144] proved to be a polypeptide of approximately 4.8 kDa, with an Fe content of 0.2% and strong catecholamine-inhibitory properties. Only a negligible amount of this material could be demonstrated in the cecum of the CV rat.[145] Baez et al. showed that α-pigment would transfer to the circulation.[142] It would seem, however, that the production of a material like

---

* The Lobund Aging Study, unpublished data.

α-pigment is not limited to cecal enlargement, but rather appears to be a result specific for the GF state. Baez et al.[146,147] report that in the (not enlarged) cecum of the GF piglet a similar norepinephrine-inhibitory material is produced, although its transfer to the bloodstream is less certain. Also of interest is the fact that the oxidation-reduction potential found in the cecum of these GF animals is again almost 300 mV more positive than in their CV counterparts.[122]

Much earlier it had been established that GF rats and mice have a considerably lower resting oxygen consumption than their CV counterparts,[29-31] although thyroid function as expressed by serum T_3 and T_4 was at least comparable (see Chapter IV). Cecectomy was found to largely abolish this difference. This led to the assumption that catecholamine-inhibitory α-pigment, via its action on the norepinephrine-controlled dissipation of metabolic energy by the brown adipose tissue (BAT), would result in a lower metabolic rate. This effect would be especially pronounced in the young GF animal with its relatively very large cecum (Figure 1). This concept seems to be corroborated by the fact that the GF rat, during the first 3 months of its existence, eats less than the CV rat, the difference becoming less with age and losing its significance after about 3 months.[148] The consequences of the above are treated in more detail in Chapter IV.

Høverstad and Midtvedt found a comparatively large amount of short-chain fatty acids in the cecum of CV rats, whereas GF rats showed only about 1% of this amount. Since similar amounts were found in the small intestine of the GF rats, they concluded that those originated from the diet. Thus, a potential source of energy, produced in the cecum of the CV animal by the fermentation of otherwise undigestible carbohydrates which may make a contribution to its energy balance, is not available in the case of the GF animal.[149] These results have been confirmed recently by Alvarez-Leite et al.,[150] who showed that even when GF rats were fed a diet containing 5% guar gum no volatile fatty acids could be detected in the cecum.

LARGE INTESTINE

Much less attention has been given to the large intestine of GF rats and mice. Whereas studies by Abram and Bishop[19] indicated that the passage of per os-administered [91]Y through the cecum of the GF mouse was greatly delayed, its passage through the colon did not seem to be much affected. As mentioned earlier, Bornstein et al. found O_2 tension somewhat lower in CV than in GF rats, whereas CO_2 tensions were higher and dependent on the composition of the microflora.[80] This appears to corroborate the more positive oxidation-reduction potential demonstrated in the cecum of the GF animal (see under Cecum), and may explain the lack of hydrogenation of dietary unsaturated fatty acids by the GF animal. GF rats excrete significantly more unsaturated fatty acids with the feces,[151] whereas GF lambs[152] contained significantly higher 18:2 and 20:4 acids in liver and fat depots. The latter would suggest that this effect also makes itself felt in the upper intestine. Bruckner and Gannoe-Hale found no difference in the neutral lipid fatty acid composition of the jejunal wall of GF and CV rats, but report substantially higher levels of 18:2 acid-containing phospholipids and lower concentrations of 20:4 acid-containing phospholipids in the GF animals, the latter being precursors to the prostaglandins and leukotrienes.[153] It has not been determined whether this PUFA-sparing effect affects dietary requirements in any substantial way.

Although substantial amounts of water are reabsorbed in the GF colon, the excreted fecal matter of the GF rat still consists of 60% or more of water, against 20 to 30% in the CV rat. This explains the loose feces observed in most GF animals. To compensate for the greater loss of water in the feces, the GF rat was found to drink 33% more water than its CV counterpart. It excreted 20% of this intake with the feces and 15% in the urine. Of its lower water intake the CV rat excreted only 5% with the feces, but 27% with the urine. Taken together, both groups excreted between 30 and 35% of their water intake in feces and urine.[87,154]

The feces of GF rats and mice, besides containing substantially higher amounts of water, also contains more free amino acids[85] and di- and polypeptides. Among those, Welling and Groen[155] describe β-aspartylglycine, one of the products apparently formed during the sterilization of the diet by autoclaving. Pelissier and Dubos[156] found another amino acid condensation product, β-aspartyl-ε-lysine, which formed upon autoclaving and, in much smaller amounts, upon irradiation. Since these products appear to withstand degradation by intestinal, but not by microbial enzymes, they may be used to check GF status.

CONCLUSIONS

As far as these GF Rodentiae and Cuniculae are concerned, cecal enlargement is the major anomaly which directly and indirectly affects many aspects of function and metabolism of the animal. Intestinal function is also altered by the effects the absence of an intestinal microflora has on the functional elements of the gut wall. In addition, the very much different composition of intestinal bile acids and the mostly higher local concentration of digestive enzymes of the GF animal all affect, to an extent, the animal's general homeostasis. Notwithstanding these differences from "normal", with proper diet and housekeeping procedures these animals thrive and can be considered useful models for those studies which ask for absence or control of the microbial factor.

REFERENCES

1. Reddy, B. S., Pleasants, J. R., Zimmerman, D. R., and Wostmann, B. S., Iron and copper utilization in rabbits as affected by diet and germfree status, *J. Nutr.*, 87, 189, 1965.
2. Yoshida, T., Pleasants, J. R., Reddy, B. S., and Wostmann, B. S., Efficiency of digestion in germfree and conventional rabbits, *Br. J. Nutr.*, 22, 723, 1968.
3. Phillips, B. P., Wolfe, P. A., and Gordon, H. A., Studies on rearing the GF guinea pig germfree, *Ann. N.Y. Acad. Sci.*, 78, 183, 1959.
4. Pleasants, J. R., Reddy, B. S., Zimmerman, D. R., Bruckner-Kardoss, E., and Wostmann, B. S., Growth, reproduction and morphology of naturally born, normally suckled germfree guinea pigs, *Z. Versuchstierk.*, 9, 195, 1967.
5. Kenworthy, R. and Allen, W. D., Influence of diet and bacteria on small intestine morphology, with special reference to early weaning and *Escherichia coli*, *J. Comp. Pathol.*, 76, 291, 1966.
6. Heneghan, J. B., Gordon, H. A., and Miniats, O. P., Intestinal mucosal surface area and goblet cells in germfree and conventional piglets, in *Clinical and Experimental Gnotobiotics. Proceedings of the VI International Symposium on Gnotobiology*, Fliedner, T. et al., Eds., Gustav Fischer, New York, 1979.
7. Reyniers, J. A., Wagner, M., Luckey, T. D., and Gordon, H. A., Survey of germfree animals: The white Wyandotte Bantam and the white Leghorn chicken, in *Lobund Reports 3*, Reyniers, J. A. et al., Eds., University of Notre Dame Press, Notre Dame, IN, 1960.
8. Salter, D. N., Nitrogen metabolism, in *The Germfree Animal in Biomedical Research*, Coates, M. E. and Gustafsson, B. E., Eds., Laboratory Animals Ltd., London, 1984, chap. 11.
9. Gustafsson, B. E., The physiological importance of the colonic microflora, *Scand. J. Gastroenterol.*, Suppl. 77, 117, 1982.
10. Ducluzeau, R., Some views on the interaction between the gastrointestinal microflora of animals and their diet, *Livest. Prod. Sci.*, 6, 243, 1979.
11. Raibaud P., Ducluzeau, R., Mocquot, G., Ghnassia, J. C., Griscelli, C., and Lauvergeon, B., Données récentes sur la flore microbienne du tube digestif chez l'homme et les animaux monogastriques, *Rev. Fr. Gynécol.*, 70, 541, 1975.
12. Wallace, J. L., Cucala, M., Mugridge, K., and Parente, L., Secretagogue-specific effects of interleukin-1 on gastric secretion, *Am. J. Physiol.*, 261, G559, 1991.
13. Hinterleitner, T. A. and Powell, D. W., Immune system control of intestinal ion transport, *Proc. Soc. Exp. Biol. Med.*, 197, 249, 1991.
14. Falus, A. and Merétey, K., Histamine: an early messenger in inflammatory and immune reactions, *Immunol. Today*, 13, 154, 1992.
15. McKay, D. M. and Bienenstock, J., The interaction between mast cells and the nerves in the gastrointestinal tract, *Immunol. Today*, 15, 533, 1994.
16. Gordon, H. A., Is the germfree animal normal? A review of anomalies in young and old age, in *The Germfree Animal in Research*, Coates, M. E., Ed., Academic Press, London, 1968, chap. 6.
17. Orland, F. J., Blayney, J. R., Harrison, R. W., Reyniers, J. A., Trexler, P. C., Wagner, M., Gordon, H. A., and Luckey, T. D., Use of the germfree animal technic in the study of experimental dental caries. I. Basic observations on rats reared free of all microorganisms, *J. Dent. Res.*, 33, 147, 1954.
18. Fitzgerald, R. J., Dental research in gnotobiotic animals, *Caries Res.*, 2, 139, 1968.
19. Abrams, G. D. and Bishop, J. E., Effect of the normal microbial flora on gastrointestinal motility, *Proc. Soc. Exp. Biol. Med.*, 126, 301, 1967.
20. Sacquet, E., Garnier, H., and Raibaud, P., Étude de la vitesse du transit gastro-intestinal des spores d'une souche thermophile stricte de *Bacillus subtilis* chez le rat "holoxénique", le rat "axénique", le rat "axénique" caecectomise, *C. R. Soc. Biol.*, 164, 532, 1970.
21. Gustafsson, B. E. and Norman, A., Influence of the diet on the turnover of bile acids in germfree and conventional rats, *Br. J. Nutr.*, 23, 429, 1969.

22. Pen, J. and Welling, G. W., Influence of the microbial flora on the amount of CCK_8- and $secretin_{21-27}$-like immunoreactivity in the intestinal tract of mice, *Comp. Biochem. Physiol.*, 76B, 585, 1983.

23. Husebye, E., Hellstrøm, P. M., and Midtvedt, T., The role of the normal microbial flora in the control of small intestine motility, in *Experimental and Clinical Gnotobiology. Proceedings of the Xth International Symposium on Gnotobiology*, Heidt, P. J. et al., Eds., *Microecol. Ther.*, 20, 389, 1990.

24. Caenepeel, P., Janssens, J., Vantrappen, G., Eyssen, H., and Coremans, G., Interdigestive myoelectric complex in germfree rats, *Dig. Dis. Sci.*, 34, 1180, 1989.

25. Reddy, B. S., Pleasants, J. R., and Wostmann, B. S., Effect of intestinal microflora on calcium, phosphorus and magnesium metabolism in rats, *J. Nutr.*, 99, 353, 1969.

26. Gordon, H. A., Bruckner-Kardoss, E., Staley, T. E., Wagner, M., and Wostmann, B. S., Characteristics of the germfree rat, *Acta Anat.*, 64, 367, 1966.

27. Wostmann, B. S., Bruckner-Kardoss, E., and Pleasants, J. R., Oxygen consumption and thyroid hormones in germfree mice fed glucose-amino acid liquid diet, *J. Nutr.*, 112, 552, 1982.

28. Wostmann, B. S., Other organs, in *The Germfree Animal in Biomedical Research*, Coates, M. E. and Gustafsson, B. E., Eds., Laboratory Animal Handbooks 9, Laboratory Animals Ltd., London, 1984, chap. X.

29. Desplaces, A., Zagury, D., and Sacquet, E., Etude de la fonction thyroidienne du rat privé de bactéries, *C. R. Acad. Sci.*, 257, 756, 1963.

30. Wostmann, B. S., Bruckner-Kardoss, E., and Knight, P. L., Cecal enlargement, cardiac output, and O_2 consumption in germfree rats, *Proc. Soc. Exp. Biol. Med.*, 128, 137, 1968.

31. Bruckner-Kardoss, E. and Wostmann, B. S., Oxygen consumption of germfree and conventional mice, *Lab. Anim. Sci.*, 28, 282, 1978.

32. Savage, D. C., Schaedler, R. W., and Dubos, R. J., Effects of bacteria on gastro-intestinal histology (abstr.), *Fed. Proc.*, 26, 803, 1968.

33. Gordon, H. A. and Bruckner-Kardoss, E., Effects of the normal microflora on various tissue elements of the small intestine, *Acta Anat.*, 44, 210, 1961.

34. Gordon, H. A. and Wostmann, B. S., Responses of the animal host to changes in environment: transition of the albino rat from germfree to the conventional state, in *Recent Progress in Microbiology*, Tunevall, G., Ed., Almquist and Wiksell, Stockholm, 1959, 336.

35. Miyakawa, M., Sumi, Y., Sakurai, K., Ukai, M., Hirabayashi, N., and Ito, G., Serum gamma globulin and lymphoid tissue in the germfree rats, *Acta Haematol. Jpn.*, 32, 501, 1969.

36. Crabbé, P. A., Nash, D. R., Bazin, H., Eyssen, H., and Heremans, J. F., Immunohistochemical observations on lymphoid tissues from conventional and germfree mice, *Lab. Invest.*, 22, 448, 1970.

37. Gordon, H. A. and Bruckner-Kardoss, E., Effects of normal microflora on intestinal surface area, *Am. J. Physiol.*, 201, 175, 1961.

38. Galjaard, H., Meer-Fiegen, W., and van der Giesen, J., Feedback control by functional villus cells on cell proliferation and maturation in the intestinal epithelium, *Exp. Cell Res.*, 73, 197, 1972.

39. Komai, M. and Kimura, S., Gastrointestinal responses to graded levels of cellulose feeding in conventional and germfree mice, *J. Nutr. Sci. Vitaminol.*, 26, 389, 1980.

40. Abrams, G. D., Microbial effects on mucosal structure and function, *Am. J. Clin. Nutr.*, 30, 1880, 1977.

41. Heneghan, J. B., Enterocyte kinetics, mucosal surface area and mucus in gnotobiotes, in *Clinical and Experimental Gnotobiotics. Proceedings of the VI International Symposium on Gnotobiology*, Fliedner, T. et al., Eds., Gustav Fischer, New York, 1979, 19.

42. Uribe, A., Alam, M., Jaramillo, E., and Midtvedt, T., Cell kinetics of the proximal jejunum epithelium in germfree rats, in *Experimental and Clinical Gnotobiology. Proceedings of the Xth International Symposium on Gnotobiology*, Heidt, P. J., Vossen, J. M., and Rusch, V. C., Eds., *Microecol. Ther.*, 20, 377, 1990.

43. Gordon, J. I., Schmidt, G. H., and Roth, K. A., Studies of intestinal stem cells using normal, chimeric, and transgenic mice, *FASEB J.*, 6, 3039, 1992.

44. Meslin, J. C. and Sacquet, E., Effects of microflora on the dimensions of enterocyte microvilli in the rat, *Reprod. Nutr. Dev.*, 24, 307, 1984.

45. Umesaki, Y., Suzuki, A., Kasama, T., Tohyama, K., Mutai, M., and Yamakawa, T., Appearance of fucolipid after conventionalization of germfree mice, *J. Biochem.*, 90, 559, 1981.

46. Meslin, J. C., Sacquet, E., and Guenet, J. L., Action de la flore bactérienne sur la morphologie et la surface de la muqueuse de l'intestin grêle du rat, *Ann. Biol. Anim. Biochim. Biophys.*, 13, 203, 1973.

47. Stevens, N.C., The Effects of the Microbial Flora on Parameters that Influence Intestinal Function. Ph.D thesis, Louisiana State University Medical Center, New Orleans, 1977.

48. Kenworthy, R. and Allen, W. D., Influence of diet and bacteria on small intestinal morphology, with special reference to early weaning and *Escherichia coli*, *J. Comp. Pathol.*, 76, 291, 1966.

49. Miniats, O. P. and Valli, V. E., The gastrointestinal tract of gnotobiotic pigs, in *Germfree Research: Biological Effects of Gnotobiotic Environments. Proceedings of the IV International Symposium on Germfree Research*, Heneghan, J. B., Ed., Academic Press, New York, 1973, 575.

50. Heneghan, J. B., Gordon, H. A., and Miniats, O. P., Intestinal mucosal surface area and goblet cells in germfree and conventional piglets, in *Clinical and Experimental Gnotobiotics. Proceedings of the VI International Symposium on Gnotobiology*, Fliedner, T. et al., Eds., Gustav Fischer, New York, 1979, 107.

51. Levevre, M. E., Joel, D. D., and Schidlovsky, G., Retention of ingested latex particles in Peyer's patches of germfree and conventional mice, *Proc. Soc. Exp. Biol. Med.*, 179, 522, 1985.

52. Knight, P. L. and Wostmann, B. S., Influence of *Salmonella typhimurium* on ileum and spleen morphology of germfree rats, *Indiana Acad. Sci.*, 74, 78, 1964.

53. Röpke, C. and Everett, N. B., Kinetics of intraepithelial lymphocytes in the small intestine of thymus-deprived and antigen-deprived mice, *Anat. Rec.*, 185, 101, 1976.

54. Crabbé, P. A., Nash, D. R., Bazin, H., Eyssen, H., and Heremans, J. F., Immunohistochemical observations on lymphoid tissues from conventional and germfree mice, *Lab. Invest.*, 22, 448, 1970.

55. Rodning, C. B., Erlandsen, S. L., Wilson, I. D., and Carpenter, A., Light microscopic morphometric analysis of rat ileal mucosa. II. Component quantitation of Paneth cells, *Anat. Rec.*, 204, 33, 1982.

56. Satoh, Y. and Vollrath, L., Quantitative electron microscopic observations on Paneth cells of germfree and ex-germfree Wistar rats, *Anat. Embryol.*, 173, 317, 1986.

57. Satoh, H., Ishikawa, K., Tanaka, H., and Ono, K., Immunohistochemical observations of immunoglobulin A in the Paneth cells of germfree and formerly-germfree rats, *Histochemistry*, 85, 197, 1986.

58. McCormack, S. A. and Johnson, L. R., Role of polyamines in gastrointestinal mucosal growth, *Am. J. Physiol.*, 260, G795, 1991.

59. McCormack, S. A., Viar, M. J., and Johnson, L. R., Polyamines are necessary for cell migration by a small intestinal crypt cell line, *Am. J. Physiol.*, 264, G367, 1993.

60. Abrams, G. D., Bauer, H., and Sprintz, H., Influence of the normal flora on mucosal morphology and cellular renewal in the ileum, *Lab. Invest.*, 12, 355, 1963.

61. Khoury, K. A., Floch, M. H., and Hersh, T., Small intestinal mucosal proliferation and bacterial flora in the conventionalization of the germfree mouse, *J. Exp. Med.*, 130, 659, 1969.

62. Guenet, J. L., Sacquet, E., Guenaeu, G., and Meslin, J. C., Action de la microflore totale du rat sur l'"activité mitotique des cryptes de Lieberkühn, *C. R. Acad. Sci.*, 270, 3087, 1969.

63. Savage, D. C., Siegel, J. E., Snellen, J. A., and Witt, D. D., Transit time of epithelial cells in the small intestines of germfree and ex-germfree mice associated with indigenous microorganisms, *Appl. Environ. Microbiol.*, 42, 996, 1981.

64. Lesher, S., Walburg, H. E., and Sacher, G. A., Generation cycle in the duodenal crypt cells of germfree and conventional mice, *Nature*, 202, 884, 1964.

65. Rolls, B. A., Turvey, A., and Coates, M. E., The influence of the gut microflora and of dietary fiber on epitheial cell migration in the chick intestine, *Br. J. Nutr.*, 39, 91, 1978.

66. Alam, M., Midtvedt, T., and Uribe, A., Cell kinetics of intestinal epithelium of germfree rats. Paper presented at the XI Int. Symp. Gnotobiology, Belo Horizonte, Brasil, 1994 (Abstr.).

67. Alam, M., Midtvedt, T., and Uribe, A., Effects of intestinal microflora on epithelial cell proliferation. Paper presented at the XI Int. Symp. Gnotobiology, Belo Horizonte, Brasil, 1994 (Abstr.).

68. Alam, M., Midtvedt, T., and Uride, A., Differential cell kinetics in the ileum and colon of germfree rats, *Gastroenterology*, 29, 445, 1994.

69. Nordström, C., Dahlqvist, A., and Josefsson, L., Quantitative determination of enzymes in different parts of the villi and crypts of rat small intestine. Comparison of alkaline phosphatase, disaccharidases and dipeptidases, *J. Histochem. Cytochem.*, 15, 713, 1968.

70. Reddy, B. S. and Wostmann, B. S., Intestinal disaccharidases in growing germfree and conventional rats, *Arch. Biochem. Biophys.*, 113, 609, 1966.

71. Olsen, W. A. and Korsmo, H. A., Sucrase metabolism in germfree rats, *Am. J. Physiol.*, 242, G650, 1982.

72. Whitt, D. D. and Savage, D. D., Kinetics of changes induced by the indigenous microbiota in the activity levels of alkaline phosphatase and disaccharidases in the small intestine enterocytes in mice, *Infect. Immun.*, 29, 144, 1980.

73. Whitt, D. D. and Savage, D. C., Influence of indigenous microbiota on the amount of protein and the activities of alkaline phosphatase and disaccharidases in the extracts of intestinal mucosa in mice, *Appl. Environ. Microbiol.*, 42, 513, 1981.

74. Yolton, D. P. and Savage, D. C., Influence of the indigenous gastrointestinal microbial flora on duodenal Mg^{2+}-dependent and $(Na^+ + K^+)$-stimulated adenosine triphosphatase activities in mice, *Infect. Immun.*, 13, 1193, 1976.

75. Miyazawa, E. and Yoshida, T., Activities of phytase and alkaline phosphatase in the intestinal contents of germfree and conventionalized rats, *Nutr. Rep. Int.*, 39, 99, 1989.

76. Corring, T., Juste, C., and Simoes-Nunes, C., Digestive enzymes in the germfree animal, *Reprod. Nutr. Dév.*, 21, 355, 1981.

77. Phillips, B. P. and Wolfe, P. A., The use of germfree guinea pigs in studies on the microbial interrelationships in amoebiasis, *Ann. N.Y. Acad. Sci.*, 78, 308, 1959.

78. Wostmann, B. S. and Bruckner-Kardoss, E., Oxidation-reduction potentials in the cecal contents of germfree and conventional rats, *Proc. Soc. Exp. Biol. Med.*, 121, 1111, 1966.

79. Eyssen, H., Piessens-Denef, M., and Parmentier, G., The role of the cecum in maintaining delta-5-steroid- and fatty acid-reducing activity of the rat intestinal microflora, *J. Nutr.*, 102, 1501, 1972.

80. Bornside, G. H., Donovan, W. E., and Myers, M. B., Intracolonic tensions of oxygen and carbon dioxide in germfree, conventional and gnotobiotic rats, *Proc. Soc. Exp. Biol. Med.*, 151, 437, 1976.

81. Ward, F. W. and Coates, M. E., Gastrointestinal pH measurements in rats: influence of microbial flora, diet and fasting, *Lab. Anim.*, 21, 216, 1987.

82. Lepkovsky, S., Wagner, M., Furuta, F., Ozone, K., and Koike, T., The proteases, amylase and lipase of the intestinal contents of germfree and conventional chickens, *Poult. Sci.*, 43, 722, 1964.

83. Reddy, B. S., Pleasants, J. R., and Wostmann, B. S., Pancreatic enzymes in germfree and conventional rats fed chemically defined, water-soluble diet free from natural substrates, *J. Nutr.*, 97, 327, 1969.

84. Genell, S., Gustafsson, B. E., and Ohlson, K., Quantitation of active pancreatic endopeptidases in the intestinal contents of germfree and conventional rats, *Scand. J. Gastroenterol.*, 11, 757, 1976.

85. Combe, E., Demarne, Y., Gueguen, L., Ivorec-Szylit, O., Meslin, J. C., and Sacquet, E., Some aspects of the relationships between gastrointestinal flora and host nutrition, *World Rev. Nutr. Diet.*, 24, 1, 1976.

86. Wostmann, B. S., Intestinal bile acids and cholesterol absorption in the germfree rat, *J. Nutr.*, 103, 982, 1973.

87. Wostmann, B. S. and Bruckner-Kardoss, E., Functional characteristics of gnotobiotic rodents, in *Recent Advances in Germfree Research. Proceedings of the VII International Symposium on Gnotobiology*, Sasaki, S. *et al.*, Eds., Tokai University Press, Tokyo, 1981, 321.

88. Bruckner, G. and Szabo, J., Nutrient absorption in gnotobiotic animals, *Adv. Nutr.*, 6, 271, 1984.

89. Kellogg, T. F. and Wostmann, B. S., Fecal neutral steroids and bile acids from germfree rats, *J. Lipid Res.*, 10, 495, 1969.

90. Madsen, D., Beaver, M. H., Chang, L., Bruckner-Kardoss, E., and Wostmann, B. S., Analysis of bile acids in conventional and germfree rats, *J. Lipid Res.*, 17, 107, 1976.

91. Riottot, M., Sacquet, E., Vila, J. P., and Leprince, C., The relationship between small intestine transit and bile acid metabolism in axenic and holoxenic rats fed different diets, *Reprod. Nutr. Dév.*, 20, 163, 1980.

92. Gustafsson, B. E. and Norman, A., Influence of the diet on the turnover of bile acids in germfree and conventional rats, *Br. J. Nutr.*, 23, 429, 1969.

93. Ranken, R., Wilson, R., and Bealmear, P. M., Increased turnover of intestinal musosal cells of germfree mice induced by cholic acid, *Proc. Soc. Exp. Biol. Med.*, 138, 270, 1971.

94. Mastromarino, A. J. and Wilson, R., Increased intestinal mucosal turnover and radiosensitivity to supralethal whole-body irradiation resulting from cholic acid-induced alterations of the intestinal microecology of germfree CFW mice, *Radiat. Res.*, 66, 393, 1976.

95. Umesaki, Y., Tohyama, K., and Mutai, M., Glycoprotein synthesis in the small intestines of germfree and conventional mice, in *Recent Advances in Germfree Research. Proceedings of the VII International Symposium on Gnotobiology*, Sasaki, S. *et al.*, Eds., Tokai University Press, Tokyo, 1981, 273.

96. Gordon, H. A. and Wostmann, B. S., Chronic mild diarrhea in germfree rodents: a model portraying host-flora synergism, in *Germfree Research. Biological Effects of Gnotobiotic Environments. Proceedings of the IV International Symposium on Germfree Research*, Heneghan, J.B., Ed., Academic Press, New York, 1973, 593.

97. Gustafsson, B. E. and Carlstedt-Duke, B., Intestinal water-soluble mucins in germfree, ex-germfree and conventional animals, *Acta Pathol. Microbiol. Immunol. Scand. Sect. B*, 92, 247, 1984.

98. Meslin, J. C. and Sacquet, E., Effects of microflora on the dimensions of enterocyte microvilli in the rat, *Reprod. Nutr. Dev.*, 24, 307, 1984.

99. Heneghan, J. B., Physiology of the alimentary tract, in *The Germfree Animal in Biomedical Research*, Coates, M. E. and Gustafsson, B. E., Eds., Laboratory Animals Handbooks 9, Laboratory Animals Ltd., London, 1984, chap. 8.

100. Heneghan, J.B., Influence of microbial flora on xylose absorption in rats and mice, *Am. J. Physiol.*, 205, 417, 1963.

101. Garnier, H. and Sacquet, E., Absorption apparente et rétention du sodium, du potassium, du calcium et du phosphore chez le rat axénique et chez le rat holoxénique, *C. R. Acad. Sci.*, 269, 379, 1969.

102. Beaver, H. M. and Wostmann, B. S., Histamine and 5-hydroxytryptamine in the intestinal tract of germfree animals, animals harbouring one microbial species and conventional animals, *Br. J. Pharmacol. Chemother.*, 19, 385, 1962.

103. Phillips, A. W., Serotonin in the germfree mouse and bacterial inhibition of intestinal serotonin, *Microecol. Ther.*, 14, 303, 1984.

104. Phillips, A., On gut membrane transport in the germfree mouse, in *Germfree Research: Microflora Control and its Application to the Biomedical Sciences. Proceedings of the VIIIth International Symposium on Germfree Research,* Wostmann, B.S., Ed., Alan R. Liss, New York, 1985.

105. Nogueira, A. A. F. M. and Barbosa, A. J. A., Immunocytochemical study of intestinal endocrine cells in germfree mice, *Eur. J. Histochem.,* 38, 213, 1994.

106. Eyssen, H., Parmentier, G., Merteus, J., and de Somer, P., The bile acids of the mouse: effects of the microflora, age and sex, in *Germfree Research: Biological Effects of Gnotobiotic Environments. Proceedings of the IV International Symposium on Germfree Research,* Heneghan, J. B., Ed., Academic Press, New York, 1973, 271.

107. Demarne, Y., Flanzy, J., and Sacquet, E., Influence of gastrointestinal flora on digestion and utilization of fatty acids in rats, in *Germfree Research: Biological Effects of Gnotobiotic Environments. Proceedings of the IV International Symposium on Germfree Research,* Heneghan, J. B., Ed., Academic Press, New York, 1973, 553.

108. Wostmann, B. S., Snyder, D. L., Johnson, M. H., and Shi, S., Functional and biochemical parameters in aging Lobund-Wistar rats, in *Dietary Restriction and Aging,* Snyder, D. L., Ed., Alan R. Liss, New York, 1989, 229.

109. Reddy, B. S., Studies on the mechanism of calcium and magnesium absorption in germfree rats, *Arch. Biochem. Biophys.,* 149, 15, 1972.

110. Andrieux, C. and Sacquet, E., Effect of microflora and lactose on the absorption of calcium, phosphorus and magnesium in the hindgut of the rat, *Reprod. Nutr. Dév.,* 23, 259, 1983.

111. Reddy, B. S., Pleasants, J. R., and Wostmann, B. S., Effect of intestinal microflora on iron and zinc metabolism, and on activities of metalloenzymes in rats, *J. Nutr.,* 102, 101, 1972.

112. Hempe, J. M. and Cousins, R. J., Cysteine-rich intestinal protein and intestinal metallothionein: An inverse relationship as a conceptual model for zinc absorption in rats, *J. Nutr.,* 122, 89, 1992.

113. Smith, J. C., McDaniel, E. G., McBean, L. D., Doft, F. S., and Halsted, J. A., Effect of microorganisms on zinc metabolism using germfree and conventional rats, *J. Nutr.,* 102, 711, 1972.

114. Wostmann, B. S., The germfree animal in nutritional studies, *Annu. Rev. Nutr.,* 1, 257, 1981.

115. Wostmann, B. S., Wong, F. R., and Snyder, D. L., Serum zinc in aging germfree and conventional rats, *Proc. Soc. Exp. Biol. Med.,* 199, 218, 1991.

116. Reddy, M. B. and Cook, J. D., Absorption of nonheme iron in ascorbic acid-deficient rats, *J. Nutr.,* 124, 882, 1994.

117. Geever, E. F., Kan, D., and Levenson, S. M., Effect of bacterial flora on iron absorption in the rat, *Gastroenterology,* 55, 690, 1968.

118. Reddy, B. S., Wostmann, B. S., and Pleasants, J. R., Iron, copper and manganese in germfree and conventional rats, *J. Nutr.,* 86, 159, 1965.

119. Andrieux, C., Gueguen, L., and Sacquet, E., Influence du mode stérilisation des aliments sur l'absorption des mineraux chez le rat axénique et holoxénique, *Ann. Nutr. Aliment.,* 33, 1257, 1979.

120. Cherian, S., Bruckner, G. G., Jackson, S., Volk, K., and Miniats, O. P., Characteristics of lower bowel contents in germfree and conventional piglets, in *Clinical and Experimental Gnotobiotics. Proceedings of the VI International Symposium on Gnotobiology,* Fliedner, T. *et al.,* Eds., Gustav Fischer, New York, 1979, 117.

121. Wostmann, B. S. and Bruckner-Kardoss, E., Development of cecal distention in germfree baby rats, *Am. J. Physiol.,* 197, 1345, 1959.

122. Kellogg, T. F. and Wostmann, B. S., Stock diet for colony production of germfree rats and mice, *Lab. Anim. Care,* 19, 812, 1969.

123. Wostmann, B. S., Pleasants, J. R., Snyder, D. L., Bos, N. A., and Benner, R., Germfree animal models: Physiology and metabolism of germfree rats and mice; their potential for further nutritional and antigenic definition, in *Laboratory Animal Studies in the Quest of Health and Knowledge*, Rothschild, H.A. *et al.*, Eds., Revista Brasiliera de Genética, 1987, 22.

124. Jervis, H. R. and Biggers, D. C., Mucosal enzymes in the cecum of conventional and germfree mice, *Anat. Rec.*, 148, 591, 1964.

125. Gustafsson, B. E. and Maunsbach, A. B., Ultrastructure of the enlarged cecum in germfree rats, *Z. Zellforsch.*, 120, 557, 1971.

126. Asano, T., Anorganic ions in cecal content of gnotobiotic rats, *Proc. Soc. Exp. Biol. Med.*, 124, 424, 1967.

127. Asano, T., Anion concentration in cecal contents of germfree and conventional mice, *Proc. Soc. Exp. Biol. Med.*, 131, 1201, 1969.

128. Nakamura, S. and Gordon, H. A., Threshold levels of NaCl upholding solute-coupled water transport in the cecum of germfree and conventional rats, *Proc. Soc. Exp. Biol. Med.*, 142, 1336, 1973.

129. Loesche, W. J. and Gordon, H. A., Water movement across the cecal wall of the germfree rat, *Proc. Soc. Exp. Biol. Med.*, 133, 1217, 1970.

130. Asano, T., Modification of cecal size in germfree rats by long-term feeding of anion exchange resin, *Am. J. Physiol.*, 217, 911, 1969.

131. Donowitz, M. and Binder, H. J., Mechanisms of fluid and electrolyte secretion in the germfree rat cecum, *Dig. Dis. Sci.*, 24, 551, 1979.

132. Carter, D., Einheber, A., Bauer, H., Rosen, H., and Burns, W. F., The role of the microflora in uremia. II, *J. Exp. Med.*, 123, 251, 1966.

133. Combe, E. and Sacquet, E., Influence de l'état axénique sur diverse composés azotés contenus dans le caecum de rats albinos recevant des quantités variables de protéines, *C. R. Acad. Sci. Paris*, 262, 685, 1966.

134. Juhr, N. C., Assoziation keimfreier Ratten mit ureolytischen Keimen, *Tierlaboratorium*, 10, 218, 1985.

135. Levenson, S. M., Crowley, L. V., Horowitz, R. E., and Malm, O. J., The metabolism of carbon-labeled urea in the germfree rat, *J. Biol. Chem.*, 234, 2061, 1959.

136. Harmon, B. G., Becker, D. E., and Jensen, A. H., Effect of microcrobial flora on nitrogen excretion products, *J. Anim. Sci.*, 26, 907, 1967.

137. Juhr, N. C. and Ladeburg, M., Intestinal accumulation of urea in germfree animals — a factor in cecal enlargement, *Lab. Anim.*, 20, 238, 1986.

138. Combe, E. and Pion, R., Note sur la composition en acides aminés du contenu de caecum de rats axéniques et de rats témoins, *Ann. Biol. Anim. Biochim. Biophys.*, 6, 255, 1966.

139. Gordon, H. A., Demonstration of a bioactive substance in caecal contents of germfree animals, *Nature*, 205, 571, 1965.

140. Gordon, H. A., A substance acting on smooth muscle in intestinal contents of germfree animals, *Ann. N.Y. Acad. Sci.*, 147, 83, 1967.

141. Baez, S. and Gordon, H. A., Tone and reactivity of vascular smooth muscle in germfree rat mesentery, *J. Exp. Med.*, 134, 846, 1971.

142. Baez, S., Bruckner, G. G., and Gordon, H. A., Responsiveness of jejunal-ileal mesentery microvessels in unoperated and cecectomized germfree rats to some smooth muscle agonists, in *Germfree Research: Biological Effects of Gnotobiotic Environments. Proceedings of the IV International Symposium on Germfree Research*, Heneghan, J. B., Ed., Academic Press, New York, 1973, 533.

143. Gordon, H. A. and Kokas, E., A bioactive pigment ("alpha pigment") in cecal contents of germfree animals, *Biochem. Pharmacol.*, 17, 2333, 1968.

144. Bruckner, G. G., Epinephrine inhibitory substance in intestinal contents of germfree rats, in *Germfree Research: Biological Effects of Gnotobiotic Environments. Proceedings of the VI International Symposium on Germfree Research*, Heneghan, J. B., Ed., Academic Press, New York, 1973, 535.

145. Wostmann, B. S., Reddy, B. S., Bruckner-Kardoss, E., Gordon, H. A., and Singh, B., Causes and possible consequences of cecal enlargement in germfree rats, in *Germfree Research: Biological Effects of Gnotobiotic Environments. Proceedings of the VI International Symposium on Germfree Research,* Heneghan, J. B., Ed., Academic Press, New York, 1973, 261.

146. Baez, S., Waldemar, Y., Bruckner, G., Miniats, O. P., and Gordon, H. A., Vascular smooth muscle depressant substance in germfree piglets, in *Clinical and Experimental Gnotobiotics. Proceedings of the VI International Symposium on Gnotobiology,* Fliedner, T. *et al.,* Eds., Gustav Fischer, New York, 1979, 129.

147. Bruckner, G., Baez, S., Miniats, O. P., and Gordon. H. A., Intestinal and vascular autonomic sensitivity of germfree and conventional piglets, in *Clinical and Experimental Gnotobiotics. Proceedings of the VI International Symposium on Gnotobiology,* Fliedner, T. *et al.,* Eds., Gustav Fischer, New York, 1979, 135.

148. Wostmann, B. S., Snyder, D. L., and Pollard, M., The diet-restricted rat as a model in aging research, *Microecol. Ther.,* 17, 31, 1987.

149. Høverstad, T. and Midtvedt, T., Short-chain fatty acids in germfree mice and rats, *J. Nutr.,* 116, 1772, 1986.

150. Alvarez-Leite, J. I., Andrieux, C., Fezerou, J., Riottot, M., and Vieira, E., Evidence for the absence of participation of the microbial flora in the hypocholestemic effect of guar gum in gnotobiotic rats, *Comp. Biochem. Physiol.,* 109A, 503, 1994

151. Eyssen, H. and Parmentier, G. G., Biohydrogenation of sterols and fatty acids by the intestinal microflora, *Am. J. Clin. Nutr.,* 27, 1329, 1974.

152. Leat, W. M. F., Kemp, P., Lyons, R. J., and Alexander, T. J. L., Fatty acid composition of depot fats from gnotobiotic lambs, *J. Agric. Sci.,* 88, 175, 1977.

153. Bruckner, G. and Gannoe-Hale, K., Fatty acid compositional changes in germfree and conventional young and old rats, in *Dietary Restriction and Aging,* Snyder, D. L., Ed., Alan R. Liss, New York, 1989, 226.

154. Wostmann, B. S., Larkin, C., Moriarty, A., and Bruckner-Kardoss, E., Dietary intake, energy metabolism, and excretory losses of adult male germfree Wistar rats, *Lab. Anim. Sci.,* 33, 46, 1983.

155. Welling, G. W. and Groen, G., β-Aspartylglycine, a substance unique to cecal contents of germfree and antibiotic treated mice, *Biochem. J.,* 175, 807, 1978.

156. Pélissier, J. P. and Dubos, F., β-Aspartyl-ε-lysine, a peptide of the fecal contents of axenic mice, *Reprod. Nutr. Dév.,* 23, 509, 1983.

Chapter IV

MORPHOLOGY AND PHYSIOLOGY, ENDOCRINOLOGY AND BIOCHEMISTRY

GENERAL ASPECTS

The usefulness of the GF animal in the study of its true genotypical potential was clearly demonstrated in a paper by du Vigneau *et al.* that appeared in 1951 in the *Journal of Nutrition.* In its summary the authors stated:

> A metabolic experiment with germfree rats fed deuterium oxide has demonstrated that animals living in the complete absence of microorganisms are capable of synthesizing the labile methyl group, as shown by the presence of deuterium in the methyl group of certain compounds such as tissue choline and creatine. The degree of synthesis was comparable to that observed in control multicontaminated rats.[1]

This first metabolic study demonstrated the usefulness of the GF animal in clearly defining its biochemical potential, and pointed to the possibility of using this model to investigate the effects of controlled microbial association on basic life processes.

The many differences in function and metabolism between GF and CV rodents express the difference between the two phenotypes, if care has been taken to keep the genotypes in as close a proximity as possible. Major causes of these differences can be found in the "enlarged cecum syndrome", and in the qualitative and quantitative differences of the challenges to the immune system. Although an insufficient number of studies are available to make a more general statement, in a number of cases cecectomy of the GF animal will bring the metabolic parameter(s) under study into or close to the CV range.

Early studies had shown that resting oxygen consumption of GF rats was significantly lower than that of their CV counterparts[2,3] (Table 1). However, as was shown in two independent studies with two different strains of rats, cecectomy brought GF oxygen consumption values at or near the values found in the CV animal.[3,4] There are good reasons to assume that materials formed in significant amounts in the GF but not in the CV cecum play a controlling role here. As early as 1965 Gordon had reported that the cecum of the GF rodent contained materials that affected smooth muscle contraction.[5] Continuing those studies, Baez and Gordon demonstrated that the microvasculature of the GF rat is quite refractory to catecholamines, vasopressin, and to a lesser extent angiotensin.[6] The material responsible for this phenomenon, termed α-pigment, appears to originate in the GF cecum and was tentatively identified be a breakdown product of ferritin. It is formed under the special conditions existing in the enlarged GF cecum, and was found to enter the circulation (see also Chapter III and below).

TABLE 1
Decrease in Metabolic Parameters of Young Adult Male Germfree Rats Expressed as Percentage of Values in Comparable Conventional Rats

Parameter	% of CV	Ref.
Oxygen uptake, Lobund-Wistar	76	3
Oxygen uptake, cecectomized	89	3
Oxygen uptake, Fischer	85	20
	76	16
Oxygen uptake, cecectomized	101	20
CO_2 output, Fischer	74	16
Cardiac output, Lobund-Wistar	68	3
Cardiac output, cecectomized	98	3
Heart weight, Lobund-Wistar	83	10
Heart weight, diet-restricted	77	10
Liver weight, Lobund-Wistar	72	10
Liver weight, diet-restricted	75	10
Thiamin per g, liver	75	33

The body of the *ad libitum*-fed rat contains a sizable amount of brown adipose tissue (BAT) which, via adrenergic stimulation, contributes significantly to the control of its energy metabolism.[7] Inhibition of the BAT-stimulating action of norepinephrine by α-pigment reduces its energy-dissipating thermogenic potential, shifting metabolic homeostasis to an increased efficiency in the use of dietary energy. This would explain the lower resting oxygen consumption of the GF rat[2,3] and the GF mouse.[8] It would also account for the lower energy intake observed in the young, growing rat[9] (Table 2) when the relative mass of the cecum reaches a maximum value of 10 to 12% of body weight at 4 to 5 weeks of age[4] (see Figure 1, Chapter III).

TABLE 2

Food Intake of Young Male Germfree and Conventional Lobund Wistar Rats Maintained *Ad Libitum* on a Casein-Starch Diet (L-474 Series[a])

Body weight	Body Weight (g/100 g)			
	Germfree	Conventional	GF/CV × 100	*p* Value
90–120	11.3	15.4	73	<0.01
160–180	8.3	10.1	82	<0.01
250–280	5.1	6.5	79	<0.01
300–330	4.5	5.1	88	0.03
350–400[b]	4.7	4.0	117	<0.05

[a]See Reddy, B. S. *et al.*[79]

[b]Nine- to 10-month-old male Lobund-Wistar rats maintained on diet L-485.[23]

Because of the above, GF rodents need less oxygen and, apparently as a consequence, have smaller cardiac outputs and smaller hearts and livers than their CV counterparts (Table 3). Although in the GF adult rat, with a relative cecal size of only 3 to 4%, the effect of the germfree state on energy requirement is eventually lost,[4] its earlier consequences on the heart, lungs, and liver appear to persist.[10] The early lower energy intake and the persistently lower oxidative exposure of the organism,[2-4,8] may then relate to the extended life span of GF rats and GF mice (see Chapter II).

TABLE 3

Comparison of Physiological and Metabolic Parameters of the Young Adult Germfree Rat With Those of Its Conventional Counterpart. Data From Assorted Publications

Body weight: same	Body temperature: –1/2 to –1°C
Food intake: less	

Percent of Conventional Value

Liver weight	79[a]
Heart weight	83[a]
Cardiac output	68[a]
Oxygen use, Wistar	74[a]
Oxygen use, Fischer	85[a]
CO_2 output, Fischer	74[a]
Serum	
T_4	116[a]
T_3	117[a]
Epinephrine	83
Norepinephrine	102

[a] Significantly different from comparable conventional value.

The effects of the strongly reduced antigenic load of the GF animal, now consisting only of the antigenic action of the diet (notably the load of sterilization-killed bacteria in materials such as commercially produced casein[11]) and bedding, is felt most directly in the development and proliferation of the cells and organs of the immune system (see Chapters VI and VII). Lately, however, it has become obvious that its cytokines have a much wider action range than only within the immune system, and will affect other organ systems as well.[12-15]

ENERGY METABOLISM

The available data are confined almost exclusively to rats. During the first 3 to 4 months the GF rat grows as well as its CV counterpart, but manages to do this with a lower food intake[9] (Table 2), lower O_2 consumption and CO_2 production,[16] and a lower cardiac output[3] (Table 1). Regional blood flow determination also indicates reduced values for the GF rat, especially for organs such as the intestine, liver, lungs, and kidneys that are directly or indirectly in contact with the microbial flora and/or its products.[17] Assuming that the oxygen consumption of the myofibrillar muscle mass of the GF rat and the CV rat are comparable, the reported difference in regional blood flow goes a long way in explaining the difference in resting oxygen consumption between GF and CV rats.

The above observations drew attention to the thyroid. The determination of thyroid function in rodents has given a somewhat confusing picture over the years, since methodology has changed over time to become ever more sophisticated. No difference in the size of the thyroid between GF and CV rats has been found.[18] Desplaces *et al.*, who first reported on the low oxygen use of the GF rat, related this now well-established phenomenon to a deficit in thyroid function, determined by ^{131}I binding, of approximately 40%.[2] However, by combining earlier T_3 and T_4 data[18-20] with the large body of serum values of male Lobund-Wistar rats available from the Lobund Aging Study (Table 4), these data do not support this conclusion except possibly for very young rats.[18] Otherwise, serum T_4 values, considered representative of the output of the thyroid, always prove to be approximately 10% higher in GF than in CV male L-W rats, and remain steady for about the first year of life. The second year of life then shows a steady decline in serum T_4 concentrations, with the GF rats still showing 10 to 15% higher values than their CV counterparts until after 18 months of age, when GF and CV values tend to become comparable (Table 4). The generally higher serum T_4 and also T_3 values of the GF rat may relate to the difference in serum TSH levels, which for the first 2 years are surprisingly stable and approximately 25% higher in GF than in CV rats.

Serum T_3 values show a slightly different picture. Although comparable in the young, fast-growing L-W males (approximately (117 ng/dl), in the GF males these values remain steady during the first year of life, only to decline during the second year. In the CV rat, T_3 values start to decline after the first early growth phase and stabilize around 90 ng/dl, a value which the GF rats only reach in the second year of life (Table 4).

TABLE 4

Serum Triiodothyronine and Thyroxine Data in Germfree
and Conventional Lobund-Wistar Rats Maintained on Diet L-485[23]

Age (months)	T_3 (ng/dl)			T_4 (µg/dl)		
	GF	CV	GF/CV	GF	CV	GF/CV
2–3	117 ± 4 (24)	117 ± 5 (18)	100	5.35 ± 0.12[a] (23)	4.61 ± 0.12 (23)	116
4–6	106 ± 6[a] (13)	91 ± 5 (19)	116	4.74 ± 0.15 (15)	4.70 ± 0.19 (20)	101
7–12	113 ± 5[a] (34)	86 ± 5 (19)	133	5.00 ± 0.19[a] (34)	4.46 ± 0.10 (45)	112
13–18	89 ± 5 (16)	96 ± 6 (14)	94	4.53 ± 0.15[a] (16)	4.03 ± 0.17 (18)	113
19–24	96 ± 6 (7)	100 ± 7 (21)	96	3.68 ± 0.16 (7)	3.89 ± 0.14 (19)	95
>24	63 ± 2 (6)	86 ± 7 (16)	77	3.64 ± 0.21 (9)	3.35 ± 0.24 (15)	109

Note: Data are averages ± SEM, taken from the Lobund Aging Study (see Chapter I).

[a] Value is significantly different from comparable CV value.

Mice appear to show a similar picture, but not enough data are available for a more definite statement.[21] After the second year no significant differences between the thyroid function indicators could be established.

Whereas the thyroid function of the GF rodent as expressed by serum T_3, T_4, and TSH concentrations is definitely not inferior to that of the CV animal, its oxygen consumption, heart size and cardiac output are less than in the CV animal. Also, its body temperature was found to be about 0.5°C lower[16,22] (Table 3). However, when we looked at O_2 uptake of liver slices and at the function of liver mitochondria, no significant difference between GF and CV rats was obvious (see under Liver).

More recently, in the Lobund Aging Study we compared heart and liver sizes of GF and CV rats that had been limited since weaning to an intake of only 12 g of diet L-485[23] per day, this being approximately 70% of the normal intake of this diet after 6 weeks of age. Here again we found similar differences, the diet-restricted GF rats showing around 75% of the heart and liver weights (per 100 g body weight) of comparable CV rats (Table 1). However, the diet-restricted GF rats now appear to reach a somewhat higher body weight than their CV counterparts.[24] A more efficient use of metabolizable energy thus appears to be a characteristic of the GF rodent. The initially low food intake of the *ad libitum*-fed GF rat[9] (Table 2) would then be a consequence inherent to the GF state, being predicated by the relatively large size of the cecum of the young animal, the site of production of the norepinephrine-inhibitory factor α-pigment (see Chapter III). α-Pigment entering the circulation would then decrease the energy-dissipating potential of BAT.

The consequently smaller size of the heart and liver, which thereafter lasts into adulthood, may eventually restrict the animal's growth to some extent since in the second year of life the *ad libitum*-fed CV rat tends to outweigh its GF counterpart. The fact that GF rats when being restricted to 70% of *ad libitum* intake always weigh slightly more than CV rats,[22] suggests that with comparable limited food intake the more efficient metabolic machinery of the GF rat appears to override the consequences of its smaller heart and liver.

However, we also found that the mature adult *ad libitum*-fed GF rat eats slightly more food than its CV counterpart.[25] Gordon reports a similar observation in GF mice.[17] In the mature rat, with its relatively much less enlarged cecum (see Figure 1, Chapter III) and presumably much lower α-pigment production, the overriding factor appears to be the loss of energy in the feces. Calculated per kilogram of body weight, 8- to 10-month-old GF rats ingest about 18% more food and 32% more water than comparable CV rats, but they excrete 88% more dry matter and about 5 times as much water with the feces.[25] Part of the dry matter consists of fibrous material which in the CV rat would have been fermented and available to the animal as small-chain fatty acids and possibly other forms of energy. In the GF animal this fermentation does not take place,[26] and chemical and caloric analysis will find all of this material in the feces. In a recent study, Juhr and Franke[27] estimated that the CV rat may ferment at least one-third of fiber and almost half of polydextroses to useful energy. In our study the amount of energy actually metabolized was 118 ± 6 and 115 ± 3 $kcal/kg^{0.75}/day$, respectively, for GF and CV rats — totally comparable values which represent 72% of GF and 81% of CV intake (Table 5). Thus, at this age increased intake appears to compensate for losses in the stool, and although no precise data on younger animals seem available, the generally recognized loose stool of GF rodents suggests that above will also hold true, at least in a qualitative way, for young animals. This fecal loss would then increase the gap in requirement for metabolizable energy between young GF and CV rats indicated by the data in Table 2.

Thus, adult GF and CV rats appear to need similar amounts of metabolizable energy,[25] although the GF rat is again able to handle this with a smaller heart and a reduced oxygen intake (Table 1). Also, when the GF rat is cecectomized early in life, the animal grows at the same rate and with food intake and oxygen use similar to those of the CV rat. Since serum T_3 and T_4 concentrations tend to be somewhat higher in the GF rat,[28] we could speculate that the catechol-inhibitory action of α-pigment would lead to a compensatory increase in thyroid function, as suggested by the consistently somewhat higher TSH levels (Table 6). Increased T_3 levels would lower the membrane potential of the hepatic mitochondria, thereby increasing oxygen consumption.[29] In the young GF rodent, with its relatively very large cecum, the inhibitory effect of its α-pigment production on energy dissipation via BAT would dominate; but at a later age

TABLE 5

Energy Balance of Adult Germfree and Conventional Male
Lobund-Wistar Rats Fed Natural Ingredient Diet L-485[23]

	Germfree (8)	Conventional (6)
Body weight, g	362 ± 12	385 ± 22
Food intake, kcal/kg/d	211 ± 9[a]	179 ± 5
kcal/kg$^{0.75}$/d	163 ± 7[a]	141 ± 3
Fecal output, kcal/kg/d	59 ± 3[a]	33 ± 2
kcal/kg$^{0.75}$/d	46 ± 2[a]	26 ± 2
Energy metabolized, kcal/kg/d	148 ± 8	143 ± 4
kcal/kg$^{0.75}$/d	116 ± 6	113 ± 3
Energy metabolized,% intake	70.9 ± 1.3[a]	80.0 ± 1.2

Note: Data are averages ± SEM, taken from Wostmann, B.S. et al.[25]
The data have been corrected for energy output in the urine.
[a] Indicates a significant difference between the groups.

TABLE 6

Serum Hormone Concentrations of Young, Young Adult, Adult, and Old Male
Germfree and Conventional Rats Maintained on Natural Ingredient Diet L-485[23]

	2–3 months		4–8 months		9–20 months		21–30 months	
	GF	CV	GF	CV	GF	CV	GF	CV
Body weight (g)	252	232	370	361	420[a]	452	444	445
	±9	±15	±4	±4	±7	±5	±13	±11
	(30)	(23)	(50)	(71)	(24)	(91)	(19)	(46)
T_4 (µg/dl)	5.35[a]	4.61	5.02[a]	4.55	4.50	4.35	3.63[a]	2.75
	±0.12	±0.12	±0.16	±0.11	±0.11	±0.10	±0.11	±0.22
	(23)	(23)	(41)	(36)	(24)	(54)	(18)	(31)
T_3 (ng/dl)	117	117	109[a]	85	99	100	75	72
	±4	±5	±5	±4	±5	±3	±8	±6
	(24)	(18)	(38)	(27)	(24)	(33)	(16)	(33)
TSH (ng/ml)	29.1[a]	18.9	23.0	20.9	29.1[a]	20.3	17.7	21.7
	±3.8	±1.3	±1.3	±1.9	±1.8	±0.7	±1.6	±1.5
	(16)	(16)	(50)	(28)	(23)	(57)	(17)	(30)
Insulin (µIU)	39.1[a]	25.5	28.9	31.1	33.4	35.8	25.0	30.1
	±2.8	±1.6	±1.8	±1.4	±2.5	±1.5	±1.5	±2.3
	(14)	(22)	(29)	(36)	(16)	(59)	(16)	(28)
Testosterone (ng/ml)	1.98	2.78	3.27	3.58	2.08	2.70	1.19	1.08
	±0.49	±0.58	±0.25	±0.47	±0.15	±0.25	±0.19	±0.20
	(12)	(20)	(43)	(32)	(24)	(46)	(18)	(33)
Prolactin (ng/ml)	164	225	301	334	305[a]	438	429	501
	±27	±58	±24	±27	±29	±34	±54	±83
	(10)	(6)	(43)	(45)	(22)	(51)	(14)	(19)

Note: Data are averages ± SEM, taken from the Lobund Aging Study (see Chapter I).
Number of observations are shown in parentheses.
[a] Indicates significant difference with comparable CV value.

this smaller effect would be compensated by thyroid hormone action,[30] thereby leading to a requirement of metabolizable energy comparable to that of its adult CV counterpart (see also Liver Function).

LIVER FUNCTION

The Lobund Aging Study confirmed earlier reports that throughout life the GF rat liver was smaller than the liver of the CV rat.[31] The same presumably is true of the GF mice,[21] although available data are not as complete as for the GF rat. In the GF rat, with a relative liver weight about 80% of the CV value,[28,31] determination of the DNA/protein ratio indicates that cellularity of the GF and CV rat liver is comparable.[28] However, its regional blood flow (ml/min/kg body weight) is only approximately 50%,[17] which creates a flow deficit per gram of liver of roughly one third. The apparently reduced requirement of the GF rat liver for oxidative metabolism is confirmed by our observation that its thiamin concentration also is only approximately 75% of that of the liver of the CV rat[32] (Table 1). In most organs thiamin expresses the availability of thiamin pyrophosphate or cocarboxylase, a major cofactor in the oxidative production of ATP. A similar situation has been reported in chicks.[33]

On the other hand, liver slices from GF rats consume oxygen at a rate comparable to that consumed by the slices from CV rats.[34] Respiration rates, respiratory control, and ADP/O values indicate that the functional integrity of mitochondria isolated from GF rats is unimpaired.[35] Although some slightly lower values were found in the GF animals, with slightly tighter respiratory control, these do not explain the difference in *in vivo* oxygen use between GF and CV rats and mice[3,21] (Table 7). Neither was any difference found in the activity of the cytochrome oxidase (EC 1.9.3.1)[36] (Table 8). It would appear that homeostatic control mechanisms, presumably influenced by the production of α-pigment in the GF cecum (see Chapter III), function to reduce the *in vivo* metabolic activity of the liver of the GF rat. Seemingly as an adaptation, the entry enzymes to the hexose monophosphate shunt (EC 1.1.1.49 and EC 1.1.1.44) show reduced activity. Of these the glucose-6-phosphate dehydrogenase is considered a "housekeeping" enzyme, adjustable by hormones, nutrients, and local oxygen availability whenever conditions require this.[37] Furthermore, the succinic acid dehydrogenase (EC 1.3.99.1), a controlling enzyme in the tricarboxylic acid cycle, also shows reduced activity (Table 8). As a result a decreased availability of cytoplasmic NADPH seems indicated.

The kidney, on the other hand, which shows only a minor deficit in regional blood flow,[16] does not show any difference in the above-mentioned dehydrogenases.[36] Although these enzymes are under the influence of the thyroid via its hormones,[38] homeostatic controls appear to override the effect of the generally higher serum T_3 and T_4 concentrations found in the GF rodents (see above). The data also suggest that the aforementioned low thiamin concentrations of the liver of the GF rat would be a result rather than a cause of the lower metabolic activity of the liver.

Table 8 brings together available data on liver enzymes. These show lower activities for the mitochondrial and cytoplasmic dehydrogenases

TABLE 7
Oxidative Energy Metablism of the Adult Male
Germfree and Conventional Fischer Rat

	Germfree	Conventional
Mitochondria[a]		
(nmol O_2/min/mg protein)	40.4 ± 3.0	41.5 ± 1.5
ADP/O	2.5	2.4
Liver slices[b]		
(nmol O_2/min/g tissue)	728 ± 41	705 ± 40
Intact rat[b]		
(ml O_2/min/kg)	17.7 ± 0.6[c]	20.7 ± 0.3

Note: Data are averages ± SEM.
[a] State 3 respiration rate; substrate β-hydroxybutyrate.
    From Sewell, D. *et al.*[35]
[b] Wostmann, B. S. and Bruckner-Kardoss, E.[20]
[c] Value different from comparable CV value, p <0.01.

TABLE 8
Metabolic Parameters and Liver Enzymes of Young Male Germfree
and Conventional Rats

	Germfree	Conventional	Ratio
Rat			
Cardiac output, ml/min/kg[a]	137 ± 7	203 ± 11	0.80[f]
O_2 consumption, ml/min/kg[a]	11.5 ± 0.4	15.2 ± 0.4	0.76[f]
O_2 consumption, ml/min/kg[b]	15.3 ± 0.4	20.2 ± 0.4	0.77[f]
CO_2 output, ml/min/kg[b]	12.3 ± 0.3	16.7 ± 0.3	0.74[f]
Rat liver			
Regional blood flow, ml/min/g[c]	2.47 ± 0.36	3.71 ± 0.28	0.67[f]
Thiamin, μg/g[d]	6.8 ± 0.1	9.1 ± 0.3	0.75[f]
Liver enzymes, units/g protein[e]			
Glucose 6-P dehydrogenase	17.8 ± 0.7	29.6 ± 2.6	0.60[f]
6-P gluconate dehydrogenase	36.4 ± 1.4	45.3 ± 2.7	0.80[f]
Succinic dehydrogenase	179 ± 2	264 ± 10	0.68[f]
Malate dehydrogenase	19.4 ± 1.1	17.0 ± 1.7	1.14
ATP-citrate lyase	12.1 ± 1.0	7.2 ± 0.5	1.68[f]
Fatty acid synthetase complex	21.4 ± 2.1	14.4 ± 1.7	1.49[f]
Cytochrome oxidase	758 ± 37	628 ± 35	1.21[g]

Note: Data are averages ± SEM, taken from the publications indicated below.
    References 3, 17, 28, and 36 pertain to Lobund-Wistar, reference 16 to
    Fischer rats.
[a] Wostmann, B. S. *et al.*[3]
[b] Calculated from Levenson, S. M. *et al.*[16]
[c] Calculated from Gordon, H. A.[17]
[d] Wostmann, B. S. *et al.*[28]
[e] Reddy, B. S. *et al.*[36]
[f] Value different from 1.00, p <0.01.
[g] Value different from 1.00, p <0.05.

in the GF rat (with the exception of the malate dehydrogenase [EC 1.1.1.40], for which no difference has been reported). However, the observation that the final oxidase in the electron transport chain shows a normal activity suggests that the potential to extract energy in the form of ATP is intact. GF rats do show a higher activity of citrate lyase (EC 4.1.3.8) and of the fatty acid synthetase complex than CV rats.[35] These do not appear to translate into major differences in body composition, which is generally comparable, although the GF rat reportedly has less body fat[39] and a slightly smaller epididymal fat pad[10] than its CV counterpart; this apparently in spite of the higher activity of the potentially fatty acid-producing enzymes.

There seems to exist a certain imbalance between dietary intake, oxygen consumption, and components of the metabolic machinery of the liver of the GF rat. Although the liver data indicate a reduced metabolic requirement, this is only translated in reduced dietary intake during the early growth period, when the potential for the production of α-pigment in the relatively greatly enlarged cecum is high, and metabolic energy production presumably is at peak efficiency (see Chapter III). Soon, however, this influence becomes less dominant, although by this time its effect on heart and lung size, and presumably on cardiac output has been established. At about 3 months of age the difference in dietary intake between the GF and the CV male Lobund-Wistar rat tends to disappear, and somewhat later the intake of the GF rat actually exceeds that of the CV animal to a certain extent, apparently to compensate for greater energy losses in the loose GF stool.[25] By the time the difference in dietary intake has waned, the persistently reduced activity of the tricarboxylic acid cycle (as suggested by the reduced activity of the succinic acid dehydrogenase and by the low thiamin content of the GF liver) and the increased activity of the cytoplasmic ATP-citrate lyase and fatty acid synthetase complex (Table 8), both enzymes positively regulated by a local increase in citric acid concentration, suggest the existence of a citric acid "escape valve" from the TCA cycle, needed because of the discrepancy between the low oxygen use and "normal" intake of dietary enegy. However, the lower activity of the entry enzymes of the hexose monophosphate shunt appears not to be in step with the increase to be expected in the demand for NADPH to convert citric acid to fatty acids and eventually triglycerides. Also, no compensatory increase was found in malate dehydrogenase, another source of NADPH. Thus, the potential excess of caloric intake suggested by the lower oxygen intake and the comparable caloric intake of the adult GF rat is not translated into body fat, but appears to escape into the blood, giving rise to an increase in citric acid in the urine.[40,41]

This would belatedly explain the needlelike calculi found in our early studies in the hearts and lungs of a strain of GF C3H mice, which upon analysis proved to consist of approximately 50% citric acid (Wostmann, B. S., unpublished data). Similar reasoning could also account for the somewhat lower body fat content of the GF rat reported by Levenson,[39] for its lower serum triglyceride levels (Table 9), and for the smaller amount of epididymal fat which becomes obvious in GF Lobund-Wistar rats after the first few months of life.

TABLE 9
Serum Glucose, Cholesterol, and Triglycerides Concentrations, and Weight of the Epidydimal Fat Pad in Young Adult and Old Germfree and Conventional Rats Maintained on Diet L-485[23]

	Germfree	Conventional	p
4–12 Months			
Glucose, mg/dl	131.0 ± 2.6 (43)	130.6 ± 2.1 (50)	NS
Cholesterol, mg/dl	85.5 ± 1.7 (43)	89.6 ± 1.9 (50)	0.10
Triglycerides, mg/dl	91.5 ± 4.1 (42)	111.1 ± 6.0 (42)	<0.01
Epididymal fat pad, mg/100 g bwt	128.8 ± 3.2 (58)	147.8 ± 3.2 (51)	<0.01
24–30 Months			
Glucose, mg/dl	158.8 ± 5.6 (10)	166.4 ± 5.1 (17)	NS
Cholesterol, mg/dl	115.2 ± 3.9 (10)	138.1 ± 5.7 (20)	0.01
Triglycerides, mg/dl	114.6 ± 12.6 (8)	148.8 ± 13.0 (19)	0.06
Epididymal fat pad, mg/100 g bwt	121.5 ± 7.7 (15)	135.0 ± 5.5 (10)	0.09

Note: Data are averages ± SEM, taken from the Lobund Aging Study. Number of animals given in parentheses.

The presence of an intestinal microflora will release large amounts of ammonia in the lower gut of the CV animal. As a result, a much higher load of ammonia will reach the liver via the portal circulation than under GF conditions.[42] In the liver, ammonia is detoxified via the urea cycle, to be eventually excreted as urea via the kidney. Studies with mice showed the activities of the urea cycle enzymes to be comparable between GF and CV animals. Only ornithine transcarbamylase (EC 2.1.3.3) was found to be more active in the livers of CV mice, obviously under the pressure of the increase input of ammonia-N via carbamoylphosphate.[43] In spite of the presumably somewhat higher activity of the overall urea cycle, no difference in BUN (blood urea nitrogen) could be established at any age between GF and CV rats in the Lobund Aging Study.

In the same vein, Katunuma *et al.* report that materials originating in the gut of the CV rat may affect the activity of tyrosine transaminase (EC 2.6.1.5).[44] Intraperitoneal administration of purified extracts of fecal cultures more than doubled the activity of this enzyme. Ziegler *et al.* came to a similar conclusion for xanthine oxidase (EC 1.2.3.2), which they found to appear earlier and reach higher levels in the livers of CV than of comparable GF mice, while no difference was found in the development of lactate dehydrogenase (EC 1.1.1.27) or glutamate-pyruvate transaminase (EC 2.6.1.2).[45] However, the differences between the xanthine oxidase levels in these mice were small, and may not contradict the earlier-mentioned lack of difference found between GF and CV rats maintained on a different type of diet. It can be concluded that, under the metabolic conditions imposed by the GF state, the liver of the GF rodent fulfills its action quite satisfactorily.

CHOLESTEROL AND BILE ACID METABOLISM

Major effects are produced by the intestinal microflora on cholesterol and bile acid metabolism which directly and/or indirectly affect many aspects of intestinal function and intermediary metabolism. Deconjugation and dehydroxylation of primary bile acids, combined with further hepatic conversion, will create totally different bile acid spectra in GF and in CV animals. This in turn will affect fecal removal of bile acids from the enterohepatic circulation, with its subsequent effect on cholesterol metabolism. Also, the microbial oxidation of cholesterol itself will affect its metabolism. Last but not least, the qualitative and quantitative differences in bile acid concentration and composition between GF and CV animals will have major effects on the functional aspects of the intestinal tract.

An excellent review of the details of bile acid metabolism in germfree animals has been given by Eyssen and van Eldere.[46]

Rats and Mice

Differences in intermediary metabolism will affect cholesterol and bile acid metabolism. The aforementioned potential paucity of cytoplasmic NADPH could favor the formation of cholesterol over fatty acids. Although this seems to be in line with the somewhat higher cholesterol concentration in the GF rat liver observed in earlier studies in which casein-starch diets were used,[47] the generally lower weight of the GF rat liver may counterbalance this effect. In at least one of our studies, total liver cholesterol and total cholesterol in the blood-liver pool showed no difference between GF and CV male Lobund-Wistar rats.[48] However, the same study did show the extent to which the absence of the "normal" intestinal microflora may otherwise influence cholesterol and bile acid metabolism. Oxidative metabolism of i.v. administered cholesterol labeled in the C^{25} position was found to be reduced by approximately one third in the GF male Wistar rat. This in turn reduces the further conversion of cholesterol to bile acids. Einarsson et al. showed that 7α-hydroxylation of

cholesterol and 12α-hydroxylation of 7α-hydroxy-4-cholesten-3-one pro-
ceed at a slower rate in liver microsomal preparations of GF than of CV
male rats.[49]

In the CV rat the resulting biliary bile acids are deconjugated and
modified by the microflora to the extent that they are less efficiently
reabsorbed in the ilium. In the CV rat the fecal loss of (largely unconjugated) bile acids is approximately twice the amount of the (conjugated,
primary) bile acids excreted by the rat in the absence of a microflora.[50]
The same holds true for the excretion of endogenous fecal neutral steroids,[51] which in the GF rat largely consist of cholesterol and some 10%
lanosterol.[52] The more complete reabsorption of conjugated bile acids
taking place in the lower GF gut results in a major increase of bile acid
in the enterohepatic circulation[53-55] (Table 10, see also Chapter III). This in
turn will slow down the conversion of cholesterol to bile acid and may
lead to an accumulation of cholesterol in the liver of the GF rat, even
when diets contain only a small amount of cholesterol.[56,57]

TABLE 10
Bile Acids in the Contents of the Third Quarter of the Small
Intestine of Male Germfree and Conventional Lobund-Wistar Rats

	Percent of Total	
	Germfree	**Conventional**
Lithocholic	—	0.9 ± 0.4
Deoxycholic	—	2.5 ± 0.4
Chenodeoxycholic	0.8 ± 0.1	0.5 ± 0.2
Cholic	51.1 ± 2.1	69.8 ± 3.2
Hyodeoxycholic	—	5.4 ± 0.8
β-Muricholic	47.9 ± 2.1	18.8 ± 1.9
ω-Muricholic	—	1.8 ± 0.3
Total (in mg)	22.2 ± 2.8	5.1 ± 1.6

Note: Data are averages ± SEM, taken from Madsen, D. et al.[55]

The resulting accumulation of cholesterol appears to have two basic
consequences for the hepatic sterol metabolism. On the one hand a compensatory reduction in the 3-hydroxy-3-methyl glutaryl (HMG)-CoA reductase (EC 1.1.1.34), the rate-controlling enzyme for cholesterol production, takes place.[58] This is in addition to the effect of the obvious absence
of bacterially produced LPS, which in the CV animal tends to stimulate
the activity of the HMG-CoA reductase.[59] On the other hand, a shift in
microsomal metabolism occurs which favors the production of chenodeoxycholic acid and its primary metabolites, the α- and β-muricholic acids, at
the expense of the production of cholic acid (Table 10). The muricholic acids
are less well reabsorbed, and thus promote bile acid secretion to relieve a
potential cholesterol accumulation. In the CV rat, β-muricholic acid is
converted by the microflora to hyodeoxycholic acid which, when reabsorbed, will be converted by the liver to ω-muricholic acid. Hyodeoxy-, and

especially ω-muricholic acid are reabsorbed only to a limited extent, thus providing an even more effective pathway for the CV rat to eliminate excess cholesterol.[60] However, even in the GF state the potential of the rat to shift from cholic acid to the production of the muricholic acids in the face of an increase in the cholesterol pool results in a relative insensitivity of its serum cholesterol to dietary cholesterol.[55,56]

Eyssen et al.[61] and Eriksson et al.[62] describe sulfatation as an important factor in murine bile acid metabolism, especially in females. Since sulfate bile acids are less well reabsorbed, this may be another way for the murine rodent to react to a potential accumulation of cholesterol in the liver and circulation.

Most of the earlier studies used relatively well-defined casein-starch diets to prevent the interference of uncontrolled dietary steroids. Casein-starch diets, however, will promote higher levels of neutral steroids.[63,64] Comparing serum and liver cholesterol values of GF rats fed either the casein-starch diet L-356[65] or a commercial formula* fortified for sterilization, we found lower serum and much lower liver values in both GF and CV male L-W rats after the animals had been transferred for only a few weeks to the commercial diet.[66] Others have established that when fed the so-called "practical-type" diets, fecal bile acid excretion was substantially higher in both GF and CV animals.[46,49] More recently in the Lobund Aging Study, when rats were maintained on the "practical-type" natural ingredient diet L-485, which does not contain any animal products,[23] serum cholesterol values of the GF rats were in the same range as found earlier on the commercial diet. With increasing age, serum cholesterol values were now somewhat lower and serum triglyceride values markedly lower in GF than in CV rats[28] (Table 9). Earlier, Wiech et al.[67] had reported the serum triglyceride concentration of Fischer rats fed casein-starch diets to be substantially lower in CV than in GF animals.

Thus, the feeding of the earlier, more defined casein-starch diets had resulted in serum cholesterol values that often were higher and liver cholesterol values that always were higher in GF animals. The feeding of natural-ingredient colony-type diets, on the other hand, appears to create conditions which outweigh those established by the more defined casein-starch diets. Feeding these diets results in comparable or possibly slightly lower serum cholesterol concentrations in younger GF rats, but definitely lower values in the older animals. Serum triglyceride concentrations are in the generally accepted range, but the GF rat values now are significantly lower than in the CV animals (Table 9). This appears to correlate with their lower fat content of body tissue.[10,39]

Far less is known about cholesterol and bile acid metabolism of mice, and not too many studies have dealt with GF mice. However, the available

---

* Rockland Rat Pellets, at the time, manufactured by A.E. Staley Manufacturing Company, Decatur, IL.

data indicate a fair resemblance to the situation in the GF rat. Again, sulfatation of bile acids seemed to play a major, though not quite well understood role.[46] In GF C3H mice β-muricholic acid production appears to develop with age — from a minor bile acid component after weaning to the dominant bile acid in the adult male, and a major component in the female.[68] As in the GF rat, intestinal bile acid concentrations were two to three times higher in the GF mouse intestine than in its CV counterpart.[20,46]

The rat's easy adaptation to increases in cholesterol intake does not make it an ideal model for the controlled study of microbial and dietary effects on cholesterol and bile acid metabolism. With its potential for the production of the muricholic acids it does not relate well to human metabolism which, lacking this ability, is much more sensitive to cholesterol intake. In this case the gerbil, and especially the gnotobiotic gerbil which, like man, produces only cholic acid and chenodeoxycholic acid as its main primary bile acids, would be a much more suitable model.

Gerbils

GF gerbils were first produced in the middle 1970s via Cesarian section and foster nursing on GF mice. However, because of greatly enlarged ceca, they never reproduced, and survivors were eventually associated with a murine-derived hexaflora consisting of *Lactobacillus brevis*, *Streptococcus faecalis*, *Enterobacter aerogenes*, *Staphylococcus epidermidis*, *Bateroides fragilis* var. *vulgatus*, and a *Fusobacterium* sp. This reduced cecal size to approximately 4% of body weight and made reproduction possible. All further studies were carried out with these gnotobiotes.[69]

As mentioned above, the gerbil does not have the rat's ability to produce the more easily eliminated muricholic acids from chenodeoxycholic acid. Its major primary bile acids are cholic acid and chenodeoxycholic acid. Although the presence of a defined hexaflora resulted in almost total deconjugation of the primary bile acid, its effect on fecal bile acid composition was less dramatic. Whereas the fecal bile acids of the true GF gerbil consisted of approximately 2/3 cholic and 1/3 chenodeoxycholic acid in conjugated form, in the hexaflora-associated animal cholic acid remained the main constituent at 90%, this higher value largely at the expense of chenodeoxycholic acid, now reduced to only 4%. Less than 2% of litho- and of deoxycholic acid was found. In contrast, the bile acids in the CV gerbil feces consisted almost 70% of deoxycholic acid, with another 20% of keto acids. A comparison of biliary and fecal bile acids in man, rat, and gerbil calculated from the data of several authors[54,69-72] is given in Table 11. As expected, even the CV gerbil, lacking the potential to produce the muricholic acids, was much more sensitive to dietary cholesterol intake than the GF or CV rat. Whereas in the GF rat a 0.10% addition of cholesterol to the diet did not result in any measurable increase in serum cholesterol,[56] in both the hexaflora-associated and the CV gerbil the effect was dramatic — 144 and 100%, respectively — with most of the increase in the VLDL and LDL fractions.[73,74] Thus, for dietary and microbiologically controlled studies of cholesterol metabolism the gnotobiotic gerbil appears to be a far better model than the CV or even the GF rat.

TABLE 11
Percentage Composition of Bile Acids in Small Intestine and Feces of Germfree and Conventional Rats, Gerbils, and Man

	Germfree				Conventional					
	Bile	Feces			Bile			Feces		
	Rat[a]	Rat[a]	Gerbil[b]	Man[c]	Rat[d]	Gerbil[b]	Man[e]	Rat[d]	Gerbil[b]	Man[f]
Lithocholic	—	—	—	—	tr[g]	11	tr	1	8	38
Deoxycholic	—	—	—	—	tr	11	26	15	68	34
Chenodeoxycholic	tr	tr	34	42	1	2	32	—	—	4
Cholic	51	48	62	52	75	73	42	4	5	3
Hyodeoxycholic	—	—	—	—	5	—	—	34	—	—
β-Muricholic	49	52	—	—	15	—	—	2	—	—
ω-Muricholic	—	—	—	—	2	—	—	19	—	—
Keto acids	tr	—	3	6	tr	3	tr	25	20	20

[a] Wostmann, B. S.[53]
[b] Wostmann, B. S., et al.[69]
[c] Kellogg, T.[70]
[d] Madsen, D. et al.[55]
[e] Angelin, B. et al.[71]
[f] Huang, C. et al.[72]
[g] tr = trace

Dogs and Pigs

Although these animals have not been consistently studied, occasional data are available. In both GF and CV dogs the major primary bile acid is cholic acid; in the GF animal overwhelmingly so (95%). The CV dog bile contains 82% cholic and 12% deoxycholic acid, the latter presumably formed via microbial action in the enterohepatic circulation. In both GF and CV dogs the biliary bile acid contains about 4% chenodeoxycholic acid. The concentration of bile acids in the bile of the GF dog is some 35% higher than in the CV dog (36.8 vs. 28.5 mg/ml), presumably because nondeconjugated bile acids of the GF dog are more easily absorbed from the lower parts of the intestine.[76]

In the pig the hyocholic acids constitute about 2/3 and chenodeoxycholic acid about 1/3 of the primary bile acids; in the GF animals hyodeoxycholic acid is only a small part, but in the CV pig a major part of that fraction,[77] again a result of microbial action in the enterohepatic circulation. The pig also produces a small amount of ω-muricholic acid. Analysis of colon contents indicates a relative increase of both hyodeoxycholic and ω-muricholic acid in both the GF and the CV animals, indicating that these bile acids are less well absorbed in the lower parts of the intestinal tract (Table 12). This suggests a potential role for these bile acids in eliminating the catabolic end products of cholesterol, especially in the CV animal.

TABLE 12
Bile Acids in Gallbladder Bile and in Lower Colon Contents of Germfree and Conventional Piglets

	Germfree		Conventional	
	Bile	Colon	Bile	Colon
Lithocholic (%)	—	—	—	14.5
Deoxycholic	—	—	—	0.7
Chenodeoxycholic	29.4	29.6[b]	25.5	0.8
Cholic	1.9[a]	5.2	0.8	1.3
Hyocholic	64.1[a]	48.1[b]	16.9	10.7
Hyodeoxycholic	4.4[a]	16.8[b]	48.1	66.0
ω-Muricholic	tr	tr[b]	0.8	3.6
Keto acids	—[a]	—[b]	7.7	2.5
Total, mg	44.3/ml	0.33/g dry[b]	39.2/ml	35.1/g dry

Note: Data from Wostmann, B. S. *et al.*[76]; tr = trace.
[a] Different from conventional bile value.
[b] Different from conventional colon content value.

OTHER INDICATORS OF FUNCTION

Although the GF state often results in quite different metabolic conditions, indicators used to check for adequacy of function, while showing certain differences between GF and CV animals, do not suggest any functional inadequacy of the liver or other organs. Studies by both Muramatsu *et al.* and Wostmann *et al.* have shown that muscle protein turnover in GF chickens[77] and GF rats[28] is comparable to that of their CV counterparts. Notwithstanding the fact that some differences exist in Zn, Fe, and Cu status, activities of mineral-dependent enzymes like alkaline phosphatase (EC 3.1.3.1), catalase (EC 1.11.1.6), and xanthine oxidase (EC 1.2.3.2) are quite comparable between GF and CV rats.[78]

Serum albumin has always been considered to be a sensitive indicator of adequate nutrition. In the present case the somewhat higher value in the GF rats suggests that their general health is as good as that of the CV animals. The slightly higher value in the GF state is a general phenomenon, explained by the lower IgG levels and the necessity to maintain the total osmolarity of the serum.

Data from blood samples obtained in the Lobund Aging Study point to a similar conclusion. Both bilirubin and BUN (blood urea nitrogen) showed no difference between GF and CV rats up to 2 years of age. Both values actually decreased with age, indicating the general good health of the animals in the study. Earlier studies by Gustafsson and Lanke[79] had established that GF rats do not reduce bilirubin to the urobilins. Uric acid levels were somewhat higher in the GF rat during the first year of life. Thereafter the difference disappeared. Creatinine values did not indicate any consistent difference between the two groups. Alkaline phosphatase was consistently lower in GF rats by some 30 to 50%. This may be in relation to the abundance of calcium which the GF rat has to handle because of increased calcium absorption in the GF state, which eventually results in a somewhat denser and heavier skeleton of the GF rat.[80] The transaminases SGOT (serum glutamic oxaloacetic transaminase) and SGPT (serum glutamic pyruvic transaminase) levels indicated little difference, the latter being slightly but significantly lower in the GF rat throughout the first 2 years of life.

WATER BALANCE AND KIDNEY FUNCTION

The loose, moist stool of the GF rodents is a fact well known to those involved in their care, since it complicates housekeeping procedures. Gordon had already shown that the dilution of intestinal material, which takes place mainly in the cecum, is only partially compensated by the water reabsorption that takes place in the colon.[17] In the absence of an actively metabolizing microflora the GF rat voids about twice as much dry matter, the difference largely consisting of high molecular weight water-retaining material. As a result the animal excretes five to six times more water per day with its feces than the CV rat. The GF rat compensates for this loss with an approximately 30% reduction of the volume of voided urine[25] (Table 13). This results in a comparable increase in the concentrations of urinary Na^+ and K^+. Cl^- concentration[81] was about twice as high as in CV urine, reflecting the somewhat higher Cl^- levels in the circulation of the GF rodent (see Mineral Metabolism below). To further compensate for fluid loss, the GF rat drinks about 35% more water than its CV counterpart. Eventually both the GF and the CV rat excrete roughly one-third of fluid intake via urine and feces.[25] Earlier data obtained from GF mice had shown an increase in water intake of 50%.[17]

No difference between GF and CV kidneys has been reported, except for the absence of D-amino acid oxidase activity reported in mice, which in CV animals appears to be triggered by D-amino acid produced by the intestinal microflora.[82] The more concentrated urine voided by the GF rodent would explain the occurrence of urinary calculi, reported especially in earlier studies when diets often contained relatively large amounts of ascorbic acid.[66,83] Reddy had found the daily Ca excretion of the GF rat, in the aforementioned somewhat smaller urinary volume, to be about twice as high as in the CV rat, presumably to compensate for the much higher Ca absorption by the GF animal.[43] Combined, these factors explain the calcium oxalate found in the urinary calculi, oxalic acid being a metabolic

TABLE 13

Water Balance of 6- to 8-Month-Old Male Germfree and Conventional Lobund-Wistar Rats Maintained on Diet L-485[23]

	Germfree	Conventional
Body weight, g	362 ± 12	385 ± 22
Intake, ml/kg/d	107.0 ± 6.2	80.7 ± 2.6
Feces, ml/kg/d	21.3 ± 2.0	3.9 ± 0.5
Urine, ml/kg/d	18.5 ± 1.9	24.6 ± 1.0
Total excreted, ml/kg/d	39.7 ± 2.1	28.5 ± 0.6
Excreted, % intake	37.1	35.3

Note: Data are averages ± SEM, taken from Wostmann, B. S. et al.[25]

end product of ascorbic acid. The substantial amount of calcium citrate found in these materials[40] (Wostmann, B.S., unpublished data) may be explained by the liver's inability to efficiently convert cytoplasmic citrate to fatty acids (see Liver Function), thereby increasing citrate excretion in the urine via the peripheral circulation.

MINERAL METABOLISM

In Chapter III the differences in intestinal uptake of a number of mineral substances have been described, among them Ca, Mg, Zn, and Fe. The data indicate generally higher rates of absorption for Ca, Mg, and possibly Zn for the GF animal,[78,80] but a diet-dependent absorption of Fe. While the GF rat maintained on the early Lobund colony diet L-462 consisting mainly of wheat flour, corn meal, and milk proteins,[84] would still show normal hemoglobin and hematocrit values,[85] similar diets resulted in severe anemia in GF but not in CV rabbits.[86] It required the replacement of the high-fiber cereals by soy meal, with relatively more organically bound Fe, to make sufficient absorption possible to ensure acceptable growth and adequate hemoglobin and hematocrit values in the GF rabbit. We assume that availability of the Fe ion, dependent on its valence state as determined by local oxidation-reduction conditions, and on phytate and other inositolphosphates present in the diet after sterilization,[87] plays a crucial role here (see Chapter III).

The strict homeostatic control of serum calcium and magnesium concentrations hardly allows for any variation in serum levels, and thereby results in the heavier skeleton of the GF rat (see Chapter III). Serum Zn values appear not to be that closely regulated. Although slightly more Zn is incorporated in the heavier skeleton of GF rats,[78] they have some 20% higher serum levels than their CV counterparts. In both GF and CV rats these levels are well maintained throughout life as long as the animal remains in a healthy condition.[88]

Na and K concentrations in the serum showed no difference between the groups, but Cl concentration were always slightly but significantly higher in the serum of the GF rat. This presumably reflects the fact that the accumulation of acid mucins in the cecum and lower gut of the GF rodent forces the small negative ions out of the gut and into the circulation (see Chapter III). This was not generally seen in the serum phosphate ion concentration, however. Values of GF and CV rats were comparable, and declined with age in a similar manner. Also comparable was the synthesis of nitrate,[89] presumably from the guanidino-N of L-arginine,[90] indicating that in this oxidative process no bacteria were involved.

ENDOCRINE SYSTEM

No dramatic differences have been reported in weight or histology of the adrenal, pituitary, thyroid, or testes. In the early days of GF research, when animals were often crowded into insufficient space, adrenal enlargement had been reported. As soon as improved housekeeping made adequate cage space available, these differences disappeared. Comparing the data of various authors, the impression is gained that in the very young GF rat the weights of the adrenal, pituitary, and testes may be slightly lower, but the data are not consistent. One report mentions lower adrenal weight in the young GF rat, possibly related to structural differences, especially in the zona reticularis of the cortex.[91] However, at around 3 to 4 months of age any significant difference seems to have disappeared[92] (Table 14).

TABLE 14

Weight of Endocrine Organs of Germfree and Conventional Young and Young Adult Male Lobund-Wistar Rats (mg/100 g Body Weight)

	Age 2–3 months		Age 6–12 months	
	Germfree	Conventional	Germfree	Conventional
Pituitary	2.2 ± 0.1[a]	3.1 ± 0.1	2.1 ± 0.1	2.1 ± 0.1
Adrenal	17.8 ± 0.6	18.2 ± 0.9	13.6 ± 0.4	12.7 ± 0.5
Thyroid[b]	6.5 ± 0.2	6.6 ± 0.2	ND	ND
Testes	889 ± 25[a]	980 ± 41	651 ± 12	667 ± 11

Note: Data are averages ± SEM, taken from the Lobund Aging Study, except where indicated; diet L-485.[23]

[a] Indicates significant difference with the adjacent CV value.

[b] Data taken from Sewell, D. L. and Wostmann, B. S.[18]

As far as the author is aware, no data are available on morphology or histology of the pancreas. In early studies Desplace *et al.* had suggested that the GF rat might be in a "prediabetic state"[2] although a later study found no evidence of impaired glucose tolerance.[93] Wiech, however, while reporting slightly elevated fasting plasma glucose levels in 2-month-old GF Fischer rats, again found glucose tolerance to be lower and insulin release during the tolerance test to be delayed in the GF animals.[67] In a later study at the Lobund Laboratory, using male Lobund-Wistar rats at 50 and 100 days of age, we were not able to detect any difference in the clearance of glucose, although the serum insulin levels obtained by the then available test methods tended to be somewhat lower in the GF rats. The same study found no difference in circulating catecholamine levels.[94] The extensive data available from the Lobund Aging Study, derived from animals maintained on the same diet as our earlier study, actually show somewhat higher serum insulin levels in the very young rats, but otherwise values are quite comparable between GF and CF rats (Table 6).

As mentioned earlier in the section on Energy Metabolism, circulating TSH levels are always some 25% higher in the GF rat, and both GF and CV rats appear to maintain constant TSH levels for at least 2 years. This results in the somewhat higher T_4 output in the GF rat during most of that time. Serum T_3 levels, being the result of diiodinization of T_4 in various organs, show a more complicated pattern. Although comparable in the very young, growing L-W males (approximately 117 ng/dl), in the GF males these values remain steady during the first year of life, and decline during the second year. In the CV rat, serum T_3 values start to decline much earlier. At the end of the fast growth period they stabilize around 90 ng/dl, a value which the GF rats only reach in the second year of life. Thus, the Lobund Aging Study shows definitely higher serum T_3 values during most of the first year of life of the GF rat, an effect which disappears during the second year. Thereafter the picture appears to be one of slow but steady decline (Tables 4 and 6). Available data suggest a similar picture for mice, but not enough data are available for a more definite statement.[21]

The only other consistent difference found between GF and CV rats, once again in the Lobund Aging Study, was in the serum prolactin values (Figure 1). On the average, serum levels in GF male rats were 25% lower than in CV rats, while in both groups concentrations climbed steadily with age. Since prolactin is known to stimulate lipid synthesis by increasing the activity of the entry enzymes to the hexose monophosphate shunt (glucose-6-phosphate dehydrogenase and 6-dehydrogenase, EC 1.1.1.49 and EC 1.1.1.44, respectively), this observation may relate to the aforementioned lower activities of these enzymes in the GF rat.[36] Because these enzymes produce the NADPH required for the formation of lipids, lower prolactin levels may thereby explain the somewhat lower fat accumulation observed in the male GF rat (see under Liver Function). Thus, one is left with the impression that subtle differences in pituitary function may exist, as expressed by higher serum TSH levels but lower prolactin concentrations, when the stimulation of a "conventional" microflora is removed from the animal.

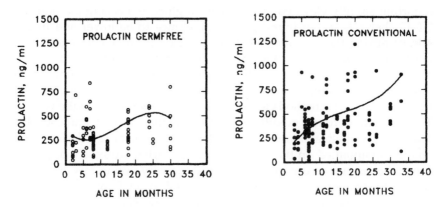

FIGURE 1

Serum prolactin concentrations of germfree and conventional male Lobund-Wistar rats. Data from the Lobund Aging Study.

The data also indicate a somewhat smaller size of the testes in the GF rat during the first 2 to 3 months of life (the Lobund Aging Study: Table 14). However, this difference disappears after 3 to 4 months. Nomura et al. report that in GF ICR-JCL mice full development of testosterone synthesis lags approximately 2 weeks behind that in CV mice[95] and is not fully active until the 8th week of life. However, data obtained in the Lobund Aging Study did not show a difference in serum testosterone concentrations even between very young GF and CV rats (Table 6). After concentrations of around 2.2 µg/dl in the first 2 to 3 months of life, the levels increased to 3.5 to 4.0 µg/dl at 8 to 9 months, and steadily declined thereafter, but with no significant difference between GF and CV Lobund-Wistar rats.

Little difference has been found between GF and CV rats as far as the development and the later regression of the thymus is concerned. In mice, where for obvious reasons the thymus has been studied more extensively, Wilson et al. report similar growth and regression, with very young GF CFW mice not quite reaching the size seen in the CV mice.[96] Data by van der Waay indicate that GF ND2 mice consistently showed lower thymus weights than their CV counterparts, at least up to 40 days of age.[97] The role of the thymus will be further elaborated in Chapter VII.

CONCLUSION

Although the concept of stress due to the "physiological inflammation" presumed to exist in the CV rat, and due at least in part to the enlarged cecum in the GF rodent, was a popular theory in the early days of GF research, presently available data do not seem to bear this out. Both GF and CV rodents appear to be well adjusted to their specific conditions and, with good management, the differences between the two do not seem to be overwhelming. A recent publication by Persson *et al.*[98] indicates that microbial activity may not be needed for the production of that ubiquitously active factor: nitric oxide. It would thus appear that notwithstanding the differences in functional parameters between GF and CV rodents, which obviously exist, presently available data leave no doubt that the GF animal model is well adjusted to its GF state, and has lost none of its capabilities to serve in a multitude of very well-controlled biomedical studies.

REFERENCES

1. du Vigneaud, V., Ressler, C., Rachele, J. R., Reyniers, J. A., and Luckey, T. D., The synthesis of "biologically labile" methyl groups in the germfree rat, *J. Nutr.*, 45, 361, 1951.
2. Desplaces, A., Zagury, D., and Sacquet, E., Étude de la fonction thyroidienne du rat privé de bactéries, *C. R. Acad. Sci.*, 257, 756, 1963.
3. Wostmann, B. S., Bruckner-Kardoss, E., and Knight, P. L., Cecal enlargement, cardiac output, and O_2 consumption in germfree rats, *Proc. Soc. Exp. Biol. Med.*, 128, 137, 1968.
4. Wostmann, B. S., Pleasants, J. R., Snyder, D. L., Bos, N. A., and Benner, R., Germfree animal models: Physiology and metabolism of germfree rats and mice: their potential for further nutritional and antigenic definition, in *Laboratory Animal Studies in the Quest of Health and Knowledge*, Rothschild, H. A. *et al.*, Eds., Revista Brasiliera de Genética, 1987, 22.
5. Gordon, H. A., Demonstration of a bioactive substance in the cecal contents of germfree animals, *Nature*, 205, 571, 1965.
6. Baez, S. and Gordon, H. A., Tone and reactivity of vascular smooth muscle in germfree rat mesentery, *J. Exp. Med.*, 134, 846, 1971.
7. Himms-Hagen, J., Brown adipose tisue thermogenesis: interdisciplinary studies, *FASEB J.*, 4, 2890, 1990.
8. Bruckner-Kardoss, E. and Wostmann, B. S., Oxygen consumption of germfree and conventional mice, *Lab. Anim. Sci.*, 28, 282, 1978.
9. Wostmann, B. S., Snyder, D. L., and Pollard, M., The diet-restricted rat as a model in aging research, *Microecol. Ther.*, 17, 31-34, 1987.
10. Snyder, D. L. and Wostmann, B. S., The design of the Lobund Aging Study and the growth and survival of the Lobund-Wistar rat, in *Dietary Restriction and Aging*, Snyder, D. L., Ed., *Prog. Clin. Biol. Res.*, 287, 39, 1989.
11. Wostmann, B. S., Pleasants, J. R., and Bealmear, P., Dietary stimulation of immune mechanisms, *Fed. Proc.*, 30, 1779, 1971.
12. Klasing, K. C., Nutritional aspects of leukocytic cytokinins, *J. Nutr.*, 118, 1436, 1988.
13. Mantovani, A., Bussolino, F., and Dejana, E., Cytokinin regulation of endothelial cell function, *FASEB J.*, 6, 2591, 1992.
14. Burke, F., Naylor, M. S., Davies, B., and Balkwill, F., The cytokine wall chart, *Immunol. Today*, 14, 165, 1993.
15. Anon., Immune-Neuroendocrinology Special Issue, *Immunol. Today*, 15, 503, 1994.
16. Levenson, S. M., Doft, F., Lev, M., and Kan, D., Influence of microorganisms on oxygen consumption, carbon dioxide production and colonic temperature of rats, *J. Nutr.*, 97, 542, 1969.

17. Gordon, H.A., Is the germfree animal normal? A review of its anomalies in young and old age, in *The Germfree Animal in Research*, Coates, M. E., Ed., Academic Press, London, 1968, 127.

18. Sewell, D. L. and Wostmann, B. S., Thyroid function and related hepatic enzymes in the germfree rat, *Metabolism*, 24, 695, 1975.

19. Ukai, M. and Mitsuma, T., Plasma triiodothyronine, thyroxine and thyrotropin levels in germfree rats, *Experientia*, 34, 1095, 1978.

20. Wostmann, B. S. and Bruckner-Kardoss, E., Functional characteristics of germfree rodents, in *Recent Advances in Germfree Research. Proceedings of the VIIth International Symposium on Gnotobiology*, Sasaki, S. *et al.*, Eds., Tokai University Press, Tokyo, 1981, 321.

21. Wostmann, B. S., Bruckner-Kardoss, E., and Pleasants, J. R ., Oxygen consumption and thyroid hormones in germfree mice fed glucose-amino acid liquid diet, *J. Nutr.*, 112, 552, 1982.

22. Kluger, M. J., Conn, C. A., Franklin, B., Freter, R., and Abrams, G. D., Effect of gastrointestinal flora on body temperature of rats and mice, *Am. J. Physiol.*, 258, R552, 1990.

23. Kellogg, T. F. and Wostmann, B. S., Stock diet for colony production of germfree rats and mice, *Lab. Anim. Care*, 19, 812, 1969.

24. Wostmann, B. S., Snyder, D. L., and Johnson, M. H., Metabolic effects of the germfree state in adult diet-restricted male rats, in *Experimental and Clinical Gnotobiology. Proceedings of the Xth International Symposium on Gnotobiology*, Heidt, P. J., Vossen, J. M., and Rusch, V. C., Eds., *Microecol. Ther.*, 20, 383, 1990.

25. Wostmann, B. S., Larkin, C., Moriarty, A., and Bruckner-Kardoss, E., Dietary intake, energy metabolism, and excretory losses of adult male germfree Wistar rats, *Lab. Anim. Sci.*, 33, 46, 1983.

26. Høverstad, T. and Midtvedt, T., Short-chain fatty acids in germfree mice and rats, *J. Nutr.*, 116, 1772, 1986.

27. Juhr, N. and Franke, J., A method for estimating the available energy of incompletely digested carbohydrates in rats, *J. Nutr.*, 122, 1425, 1992.

28. Wostmann, B. S., Snyder, D. L., Johnson, M. H., and Shi, S., Functional and biochemical parameters in aging Lobund-Wistar rats, in *Dietary Restriction and Aging*, Snyder, D. L., Ed., *Prog. Clin. Biol. Res.*, 287, 229, 1989.

29. Gregory, R. B. and Berry, M. N., The administration of triiodothyronine to rats results in a lowering of the mitochondrial membrane potential in isolated hepatocytes, *Biochim. Biophys. Acta*, 1133, 89, 1991.

30. Houstec, J., Kopecky, J., Baudysova, M., Janikova, D., Pavelka, S., and Klement, P., Differentiation of brown adipose tissue and biogenesis of thermogenic mitochondria *in situ* and in cell culture, *Biochim. Biophys. Acta*, 1018, 243, 1990.

31. Gordon, H. A., Bruckner-Kardoss, E., Staley, T. E., Wagner, M., and Wostmann, B. S., Characteristics of the germfree rat, *Acta Anat.*, 64, 301, 1966.

32. Wostmann, B. S., Knight, P. L., Keeley, L. L., and Kan, D. F., Metabolism and function of thiamine and naphthoquinones in germfree and conventional rats, *Fed. Proc.*, 22, 120, 1963.

33. Coates, M. E., Ford, J. F., and Harrison, G. F., Intestinal synthesis of vitamins of the B complex in chicks, *Br. J. Nutr.*, 22, 493, 1968.

34. Wostmann, B. S., Reddy, B. S., Bruckner-Kardoss, E., Gordon, H. A., and Singh, B., Causes and possible consequences of cecal enlargement in germfree rats, in *Germfree Research. Biological Effects of Gnotobiotic Environments. Proceedings of the IV International Symposium on Germfree Research*, Heneghan, J. B., Ed., Academic Press, New York, 1973, 261.

35. Sewell, D. L., Wostmann, B. S., Gariola, C., and Aleem, M. I. H., Oxidative energy metabolism in germfree and conventional rat liver mitichondria, *Am. J. Physiol.*, 228, 526, 1975.

36. Reddy, B. S., Pleasants, J. R., and Wostmann, B. S., Metabolic enzymes in the liver and kidney of the germfree rat, *Biochim. Biophys. Acta*, 320, 1, 1973.
37. Kletzien, R. F., Harris, P. K. W., and Foellmi, L. A., Glucose-6-phosphate dehydrogenase: a "housekeeping" enzyme subject to tissue-specific regulation by hormones, nutrients, and oxidant stress, *FASEB J.*, 8, 174, 1994.
38. Rivlin, R. S., Regulation of flavoprotein enzymes in hypothyroidism and in riboflavin deficiency, in *Advances in Enzyme Regulation*, Vol. 8, Weber, G., Ed., Pergamon Press, New York, 1970, 239.
39. Levenson, S. M., The influence of the indigenous microflora on mammalian metabolism and nutrition, *J. Parent. Ent. Nutr.*, 2, 75, 1978.
40. Glas, J. E. and Gustafsson, B. E., Mineral pattern of urinary calculi from germfree rats, *Acta Radiol.*, 1, 363, 1963.
41. Gustafsson, B. E. and Norman, A., Urinary calculi in germfree rats, *J. Exp. Med.*, 116, 273-284, 1963.
42. Warren, K. S. and Newton W. L., Portal and peripheral blood ammonia concentrations in germfree and conventional guinea pigs, *Am. J. Physiol.*, 197, 717, 1959.
43. Saheki, T., Ueda, A., Hosoya, M., Kasunuma, T., Ohnishi, N., and Ozawa, A., Comparison of the urea cycle in conventional and germfree mice, *J. Biochem. (Tokyo)*, 88, 1563, 1980.
44. Katunuma, N., Kominami, E., and Tomino, I., Enteric flora and mammalian enzyme induction, in *Advances in Enzyme Regulations*, Vol. 9, Weber, G., Ed., Pergamon Press, New York, 1971, 291.
45. Ziegler, R. W., Hulchinson, H. D., and Hegner, J. R., A comparison of xanthine oxidase, lactate dehydrogenase, and glutamate-pyruvate transaminase activities in germfree and conventional mice, *Int. J. Biochem.*, 1, 349, 1970.
46. Eyssen, H. and van Eldere, J., Metabolism of bile acids, in *The Germfree Animal in Biomedical Research*, Coates, M. E. and Gustafsson, B. E., Eds., Laboratory Animal Handbook 9, Laboratory Animals Ltd., London, 1984, 291.
47. Wostmann, B. S. and Wiech, N. L., Total serum and liver cholesterol in germfree and conventional male rats, *Am. J. Physiol.*, 201, 1027, 1961.
48. Wostmann, B. S., Wiech, N. L., and Kung, E., Catabolism and elimination of cholesterol in germfree rats, *J. Lipid Res.*, 7, 77, 1966.
49. Einarsson, K., Gustafsson, J. A., and Gustafsson, B. E., Differences in germfree and conventional rats in liver microsomal metabolism of steroids, *J. Biol. Chem.*, 218, 3623, 1973.
50. Gustafsson, B. E. and Norman, A., Influence of the diet on the composition of fecal bile acids in rats, *Br. J. Nutr.*, 23, 627, 1969.
51. Kellogg, T. F., Knight, P. L., and Wostmann, B. S., Effect of bile acid deconjugation on the fecal excretion of steroids, *J. Lipid Res.*, 11, 362, 1970.
52. Kellogg, T. F. and Wostmann, B. S., Fecal neutral steroids and bile acids from germfree rats, *J. Lipid Res.*, 10, 495, 1969.
53. Wostmann, B. S., Intestinal bile acids and cholesterol absorbtion in the germfree rat, *J. Nutr.*, 103, 982, 1973.
54. Sacquet, E., van Heijenoort, Y., Riottot, M., and Leprince, C., Action de la flore microbienne du tractus digestif sur le metabolism des acides bilaires chez le rat, *Biochim. Biophys. Acta*, 380, 52, 1975.
55. Madsen, D., Beaver, M., Chang, E., Bruckner-Kardoss, E., and Wostmann, B. S., Analysis of bile acids in conventional and germfree rats, *J. Lipid Res.*, 17, 107, 1976.
56. Kellogg, T. F. and Wostmann, B. S., The response of germfree rats to dietary cholesterol, in *Germfree Biology*, Mirand, E. A. and Back, N., Eds., Plenum Press, New York, 1969, 293.
57. Ukai, M., Tomura, A., and Ito, M., Cholesterol synthesis in germfree and conventional rats, *J. Nutr.*, 106, 1175, 1976.

58. Einarson, K., Gustafsson, J. A., and Gustafsson, B. E., Hepatic 3-hydroxy-3-methylglutaryl CoA reductase activity in germfree rats, *Proc. Soc. Exp. Biol. Med.,* 154, 319, 1977.
59. Feingold, K. R., Hardardottir, I., Mennon, R., Krul, E. J. T., Moser, A. H., Taylor, J. M., and Grunfeld, C., Effect of endotoxin on cholesterol biosynthesis and distribution in serum lipoproteins in Syrian hamsters, *J. Lipid Res.,* 34, 2147, 1993.
60. Wostmann, B. S., Beaver, M. H., Chang, L., and Madsen, D. C., Effect of autoclaving a lactose-containing diet on cholesterol and bile acid metabolism of conventional and germfree rats, *Am. J. Clin. Nutr.,* 30, 1999, 1977.
61. Eyssen, H., Smets, L., Parmentier, G., and Janssen, G., Sex-linked differences in bile acid metabolism of germfree rats, *Life Sci.,* 21, 707, 1977.
62. Eriksson, H., Taylor, W., and Sjövall, J., Occurence of sulfated 5α-cholanoates in rat bile, *J. Lipid Res.,* 19, 177, 1978.
63. Scholz-Ahrens, K. E., Hagemeister, H., Unshelm, J., Agergaard, N., and Barth, C., Response of hormones modulating plasma cholesterol to dietary casein or soy protein in minipigs, *J. Nutr.,* 120, 1387, 1990.
64. De Schrijver, R., Cholesterol metabolism in mature and immature rats fed animal and plant proteins, *J. Nutr.,* 120, 1624, 1990.
65. Larner, J. and Gillespie, R. E., Gastrointestinal digestion of starch. III. Intestinal carbohydrase activities in germfree and non-germfree animals, *J. Biol. Chem.,* 225, 279, 1957.
66. Wostmann, B. S. and Kan, D., The cholesterol-lowering effect of commercial diet fed to germfree and conventional rats, *J. Nutr.,* 84, 277, 1964.
67. Wiech, N. L., Hamilton, J. G., and Miller, O. N., Absorption and metabolism of dietary triglycerides in germfree and conventional rats, *J. Nutr.,* 93, 324, 1967.
68. Eyssen, H., Parmentier, P., Mertens, J., and de Somer, P., The bile acids of the mouse: effect of microflora, age and sex, in *Germfree Research: Biological Effect of Gnotobiotic Environments. Proceedings of the IVth International Symposium on Germfree Research,* Heneghan, J. B., Ed., Academic Press, New York, 1973, 271.
69. Wostmann, B. S., Beaver, M., Bartizal, K., and Madsen, D., Gnotobiotic gerbils, in Proc. 4th Int. Symp. Contamination Control, Vol. 4, Washington, D.C., 1978, 132.
70. Kellogg, T. F., Fecal bile acids and neutral sterols of gnotobiotic, antibiotic fed normal, and normal human children, in *Germfree Research. Biological Effects of Gnotobiotic Environments. Proceedings of the IV International Symposium on Germfree Research,* Heneghan, J. B., Ed., Academic Press, New York, 1973, 79.
71. Angelin, B., Einarsson, K., and Helstrom, K., Evidence for the absorption of bile acids in the proximal small intestine of normo- and hyperlipidaemic subjects, *Gut,* 17, 420, 1976.
72. Huang, C. T., Rodriguez, J. T., Woodward, W. E., and Nichols, B. L., Comparison of patterns of fecal bile acids and neutral sterols between children and adults, *Am. J. Clin. Nutr.,* 29, 1196, 1976.
73. Bartizal, K. F., Beaver, M. H., and Wostmann, B. S., Cholesterol metabolism in germfree gerbils, in *Recent Advances in Germfree Research. Proceedings of the VII International Symposium on Gnotobiology,* Sasaki, S. *et al.,* Eds., Tokai University Press, Tokyo, 1981, 269.
74. Bartizal, K. F., Beaver, M. H., and Wostmann, B. S., Cholesterol metabolism in gnotobiotic gerbils, *Lipids,* 17, 791, 1982.
75. Beaver, M. H., Wostmann, B. S., and Madsen, D. C., Bile acids in bile of germfree and conventional dogs, *Proc. Soc. Exp. Biol. Med.,* 157, 386, 1978.
76. Wostmann, B. S., Beaver, M., and Madsen, D., Bile acid in germfree piglets, in *Clinical and Experimental Gnotobiotics. Proceedings of the VI International Symposium on Gnotobiology,* Fliedner, T. *et al.,* Eds., Gustav Fischer, New York, 1979, 121.
77. Muramatsu, T., Salter, D.N., and Coates, M.E., Protein turnover in breast muscle of germfree and conventional chicks, *Br. J. Nutr.,* 54, 131, 1985.

78. Reddy, B. S., Pleasants, J. R., and Wostmann, B. S., Effect of intestinal microflora on iron and zinc metabolism, and on the activities of metalloenzymes in rats, *J. Nutr.,* 102, 101, 1972.

79. Gustafsson, B. E. and Lanke, L. S., Bilirubin and urobilins in germfree, ex-germfree, and conventional rats, *J. Exp. Med.,* 112, 975, 1960.

80. Reddy, B. S., Pleasants, J. R., and Wostmann, B. S., Effect of intestinal microflora on calcium, phosphorus and magnesium metabolism in rats, *J. Nutr.,* 99, 353, 1969.

81. Lev, M., Alexander, R. H., and Levenson, S., Impaired water metabolism in germfree rats, *Proc. Soc . Exp. Biol. Med.,* 135, 700, 1970.

82. Lyle, L. R. and Jutila, J. W., D-amino acid oxidase induction in the kidneys of germfree mice, *J. Bacteriol.,* 96, 606, 1968.

83. Gustafsson, B. E., Lightweight stainless steel systems for rearing germfree animals, *Ann. N.Y. Acad. Sci.,* 78, 17, 1959.

84. Wostmann, B. S., Nutrition of the germfree mammal, *N.Y. Ann. Sci.,* 78, 175, 1959.

85. Reddy, B. S., Wostmann, B. S., and Pleasants, J. R., Iron, copper and manganese in germfree and conventional rats, *J. Nutr.,* 86, 159, 1965.

86. Reddy, B. S., Pleasants, J. R., Zimmerman, D. L., and Wostmann, B. S., Iron and copper utilization in rabbits as affected by diet and germfree status, *J. Nutr.,* 87, 189, 1965.

87. Brune, M., Rossander-Hultén, L., Hallberg, L., Gleerup, A., and Sandberg, A., Iron absorption from bread in humans: Inhibiting effects of cereal fiber, phytate and inositol phosphates with different numbers of phosphate groups, *J. Nutr.,* 122, 442, 1992.

88. Wostmann, B. S., Wong, F. R., and Snyder, D. L., Serum zinc in aging germfree and conventional rats, *Proc. Soc. Exp. Biol. Med.,* 199, 218, 1992.

89. Green, L. C., Tannenbaum, S. R., and Goldman, P., Nitrate synthesis in germfree and conventional rats, *Science,* 212, 56, 1981.

90. Nathan, C., Nitric oxide as a secretory product of mammalian cells, *FASEB J.,* 6, 3051, 1992.

91. Wakabayashi, T., Takahashi, T., and Miyakawa, M., Histochemical and electron microscopic studies of the adrenal cortex of germfree rats, in *Advances in Germfree Research and Gnotobiology,* Miyakawa, M. and Luckey, T. D., Eds., CRC Press, Boca Raton, FL, 1968, 114.

92. Wostmann, B. S., Other organs, in *The Germfree Animal in Biomedical Research,* Coates, M. E. and Gustafsson, B. E., Eds., Laboratory Animal Handbook 9, Laboratory Animals Ltd. London, 1984, 215.

93. Desplaces, A., Zagury, D., and Saquet, E., Épreuve d'hyperglycémie provoquée chez le rat; étude comparative du rat normal et du rat 'germfree', *C. R. Acad. Sci. Paris,* 260, 4821, 1965.

94. Sewell, D. L., Bruckner-Kardoss, E., Lorenz, L., and Wostmann, B. S., Glucose tolerance, insulin and catecholamine levels in germfree rats, *Proc. Soc. Exp. Biol. Med.,* 152, 16, 1978.

95. Nomura, T., Ohshawa, N., Kageyama, K., Saito, M., and Tajima, Y., Testicular functions of germfree mice, in *Germfree Research: Biological Effects of Gnotobiotic Environments. Proceedings of the IV International Symposium on Germfree Research,* Heneghan, J. B., Ed., Academic Press, New York, 1973, 515.

96. Wilson, R., Bealmear, P. M., and Sobonya, R., Growth and regression of the germfree (axenic) thymus, *Proc. Soc. Exp. Biol. Med.,* 118, 97, 1965.

97. Van der Waay, D., Influence of intestinal flora on the relative thymus weight, in *Gnotobiology and Its Applications. Proceeding of the IX International Symposium on Gnotobiology,* Ed. Fond. Marcel Merieux, Lyon, France, 1987, 281.

98. Persson, M. G., Midtvedt, T., Leone, A. M., and Gustafsson, L. E., Ca^{2+}-dependent and Ca^{2+}-independent exhaled nitric oxide, presence in germfree animals, and inhibition by arginine analogues, *Eur. J. Pharmacol.,* 264, 13, 1994.

Chapter V

NUTRITION

GENERAL ASPECTS

The nutritional requirements of a GF animal are determined first of all by the phenotypic requirements of the metabolic machinery existing in the GF state. The following factors may affect the requirements of the animal as a whole:

1. The physiological status of the intestinal wall in the absence of an intestinal microflora.
2. The effect of the absence of a microflora on the digestive elements in the lumen of the intestinal tract (bile acids, enzymes, see Chapter III).
3. Effects of the absence of the microflora on the physical conditions in the tract (pH, oxidation-reduction potential).
4. Absence of the microbial synthesis of essential nutrients which might otherwise be available to the host, as exemplified by the phylloquinones, certain B vitamins, and others.
5. Absence of microbial binding or possible destruction of nutrients, including the microbial chelating potential for certain minerals.

Presently available data indicate that the metabolic requirements of the GF rat and the GF mouse differ little from those of their CV counterparts, except for their somewhat lower energy requirements and apparently lower requirements of certain amino acids. As discussed in Chapter IV, the GF rat and the GF mouse appear to have a more efficient energy metabolism, especially when young, presumably related to the catecholamine-inhibitory materials originating in their large cecal masses. This is demonstrated by their lower food intake during early growth, their lower O_2 consumption, and lower thiamin levels in the liver — thiamin pyrophosphate being a cofactor for the functioning of the pyruvate dehydrogenase, the transketolase, and other essential metabolic enzymes. However, with the exception of the consequences of this syndrome, which is expressed in the changes of the activity of specific hepatic enzymes (see also Chapter IV), their biochemical machinery with its hormonal controls appears to be rather similar to that of the CV animal. When restricted to

71

only 12 g of diet per day, GF Lobund-Wistar (L-W) rats are able to maintain a somewhat higher body weight than their CV counterparts, again indicating more efficient metabolism (see Chapter IV).

The early investigators had none of this information available and had to work with the rather sketchy knowledge of the requirements of the CV animal available at that time. Especially at the end of the 19th and the beginning of the 20th century, this severely limited the survival of the first GF guinea pigs[1] and GF chickens.[2] However, Cohendy and Wollman had already observed that when GF chickens were associated at hatching with *Escherichia coli* their survival was substantially prolonged, indicating an as yet unknown beneficial effect of the presence of this microorganism.[2] During the 1930s Glimstedt[3] was able to extend the life span of the GF guinea pig beyond 2 months by adding known sources of vitamins and fruit juice to the diet; but even then, as he described the underdeveloped lymphoid tissue of the GF animal, doubt always existed as to whether this deficit was caused, at least in part, by the stress of inadequate nutrition.

It was in the 1940s and 1950s that successful rearing of GF rats, mice, and guinea pigs was reported.[4-7] At that time every possible source of nutrients was incorporated in the diet in an effort to provide, after sterilization, the nutritional components apparently required by the GF state. Further dietary developments made it possible to start reproducing GF rat colonies in the late 1940s (see below). In 1959 Phillips *et al.* reported the conditions necessary to maintain GF guinea pigs for longer than one year, although the animals never reproduced.[8] Reproduction was reported in 1969, after further diet development.[9] It was only after Pleasants *et al.*[10] showed that GF BALB/c mice could be maintained through eight generations on a diet consisting of well-defined small molecular chemical entities that it could be assumed the metabolic requirements of that mouse strain had been, at least qualitatively, established.

In 1987 the International Council for Laboratory Animal Science (ICLAS) published the *ICLAS Guidelines on the Selection and Formulation of Diets for Animals in Biomedical Research*.[11] This monograph contains the latest considerations on animal nutrition in general, practicalities about diet composition, dietary ingredients, the sterilization of diets, and also a short description of the most pertinent facts about the development and use of the chemically defined, antigen-free, water-soluble diet formulas. It is highly recommended as a very practical guide for those using experimental animals.

In the following we will trace the development of the various presently used rat and mouse diets, since this depicts the specific problems which had to be overcome. GF gerbils did well on the colony diet developed for GF rats and mice after about 35 to 40% had been replaced by Quaker Oats. This addition has been found to reduce the size of the cecum in both GF and GN gerbils and GF guinea pigs (for details see Chaper IV, References 73 and 74). Specifics on diets for GF rabbits and GF guinea pigs are found in the above-mentioned ICLAS publication and in pertinent publications cited in Chapters III and IV.

EARLY RAT AND MOUSE DIETS

Hand-feeding of the Cesarian-Derived Newborn

Since GF rats and GF mice are presently available commercially and are maintained at many scientific institutions, the reader is referred to papers by Pleasants[6,12] for a description of this very special, but for our purpose here inexpedient, subject matter. However, the experience gained in hand-feeding soon led to the development of the chemically defined, water-soluble amino acid-dextrose diets that are described in Chapter VI.

GF gerbils were obtained only by foster-nursing on GF mice.[13]

Solid Diets (Table 1)

TABLE 1
Early Diets for Germfree Rats (g/100 g)

L-356[a]		L-462[b]	
Casein, Labco	20	Casein, Sheffield	5
Rice flour	58	Lactalbumin	10.5
Cellulose spangles	5	Milk powder, whole	10.5
Corn oil	5	Wheat flour, whole	32
Yeast extract	2	Corn meal, yellow	34
Liver powder	2	Liver powder	2
		Alfalfa	2
Vitamins		Vitamins	
Salts		Salts	
Water added	10	Water added	15
N × 6.25	22	N × 6.25	24
Fat	8	Fat	8
Ash	5.4	Ash	3.5

[a] This diet was originally designed by Larner and Gillespie.[14] A more detailed description of a later, slightly modified form can be found in diet L-464.[40]
[b] For further details see Wostmann.[15]

At the Lobund Laboratory, colony production of GF rats and GF mice of satisfactory quality started with the use of diet L-356.[14] This diet was based on rice flour, casein, corn oil, and vitamin and mineral supplements judged adequate at the time. Keeping the above-mentioned experience of Cohendy in mind, 2% liver powder and 2% yeast extract were added to provide for any possibly unknown nutrients otherwise provided by intestinal microorganisms. Since this was an expensive diet for colony production, a few years later a more practical and less expensive diet, L-462, was developed that was based on wheat flour, corn meal, and whole milk powder, with added lactalbumin, casein, and vitamin and mineral supplements.[15] Again, 2% each of alfalfa meal and liver powder was added as a safeguard. This diet proved to be quite satisfactory, especially since it reduced the occurrence of soft tissue calcification, later ascribed to the presence of rather substantial amounts of ascorbic acid in earlier diets. In Sweden, Gustafsson reported consistent colony production and good quality animals on a simpler and already better-defined diet consisting of wheat starch, casein, and arachis oil, with added vitamin and mineral supplements.[16]

FURTHER DEVELOPMENT OF RAT AND MOUSE DIETS

Originally, the aforementioned diets had been developed for the GF rat; but in the early 1950s the GF Swiss-Webster mouse joined the group of GF rodents,[6] and although there have been some suggestions that different strains of mice have different nutritional requirements, it would appear that, in general, diets satisfactory for the GF rat will work well for the GF mouse.

Further diet development was driven by a number of factors. Although satisfactory growth, reproduction, life span, and the absence of pathological conditions had to remain the main criteria, cost soon became a factor. Most early diets used casein, which became a serious budget item as soon as animal production expanded. This led to the cheaper, general purpose diets as exemplified by L-485 (Table 2)[17] and Ralston Purina's formula 5010 C,* composed of cheaper materials with supplements added to provide satisfactory performance after sterilization at 122°C for 20 min. Diet L-485 contains only soy and corn products, with added alfalfa, as sources of protein. It has been used for more than 25 years at the Lobund Laboratory for colony production of GF rats and GF mice, and supports excellent growth, reproduction, and life span.

However, there soon developed a mounting interest in nutritional and other studies which required as much definition of dietary intake as possible. Consequently there was a demand for diets in which each dietary component could be manipulated independently, e.g., in which casein provided only protein and with no additional small amounts of fat-soluble vitamins. This resulted in experimental diets generally consisting of fat-extracted casein fortified with methionine, a well-defined starch, 5 to 10% of a vegetable oil, and with added vitamin and mineral supplements. Examples are Gustafsson's early diet D7,[16] the diet developed by Sacquet's group in France,[18] and the Lobund L-474 series.[19] These formulas could

* Ralston-Purina Co., St. Louis, MO.

TABLE 2
Composition of Natural Ingredient Diet L-485[17]

Ingredients	g/kg	Nutrient Composition (%)	
Ground corn	590	Protein	20.0
Soy bean meal, 50% CP[a]	300	Fat	5.3
Alfalfa meal, 17% CP	35	Fiber	3.0
Corn oil	30	Ash	5.5
Iodized NaCl	10	Moisture	11.2
Dicalcium phosphate	10	Nitrogen-free material	55.0
Calcium carbonate	5		
Lysine (feed grade)	5	Gross energy 3.9 kcal/g	
Methionine (feed grade)	5		
Vitamin and mineral mixes	0.25[b]		
BTH	0.125		

[a] Crude protein.
[b] See Kellogg, T. F. and Wostmann, B. S.[17]

be modified according to experimental needs, but their basic compositions remained and still remains the same, except that increased knowledge of vitamin and mineral requirements have led to some revision of supplements. Well-defined diets for specific experimental purposes were developed on this basis, as is exemplified by Lobund's diet L-488F[20] which is based on the aforementioned L-474 diets but contains an additional 0.05% cholesterol, corresponding to a human intake of between 250 and 300 mg cholesterol per day (Table 3).

TABLE 3
Composition of Casein-Starch Diet L-488F
Fortified With Cholesterol (g/100 g)

Casein	24.0
DL-Methionine	0.3
Starch	60.4
Cellophane spangles	5.0
Corn oil[a]	5.0
Cholesterol	0.05
Vitamin and mineral mixes[b]	5.3

[a] Contains fat-soluble vitamins.
[b] For details of the vitamin and mineral mixes
see Reddy et al., Chapter III, Reference 83.

These more defined diets all allowed satisfactory growth and freedom from overt pathogenicity for the duration of the experiment. However, the necessary sterilization, even when done by gamma radiation, always produced a certain amount of rather uncontrolled change in dietary components besides adding to the often substantial antigenicity of the diet. Absolute control of dietary intake would come only with the development of the chemically defined, water-soluble diets that could be sterilized by filtration. Besides making it possible to establish qualitative and possibly quantitative requirements, these diets also provide a minimum of dietary antigenicity (see Chapter VI).

Starting in the 1970s, however, animals in colony production and those used in long-term studies were generally maintained on the afore-mentioned practical type diets. These now support good growth and reproduction and a satisfactory life span, especially since they generally do not contain casein with its potential to cause kidney pathology.[21]

STERILIZATION OF SOLID DIETS

All diets for GF animals obviously need to be absolutely sterile. Here we will briefly summarize the most salient points pertaining to steriliza-tion procedures of solid diets.

In most instances sterilization is done with live steam. Under ideal circumstances a vacuum is drawn before the entry of steam into the autoclave. This allows quick entry of the steam into the diet mass (pref-erably pellets), and should bring the core of the diet up to 120 to 122°C within minutes. This temperature should then be maintained for at least 15 min. Thereafter the steam lines are closed, and a second vacuum is drawn to remove moisture and to lower the temperature in the autoclave as quickly as possible to avoid unnecessary nutrient loss of the already sterile diet. This procedure will ensure uniformity of operation, indepen-dent of the equipment used. The resulting diet will have lost some of its protein quality due largely to diminished availability of lysine, a substan-tial loss of thiamin, and smaller losses of other B vitamins.[22] These must be compensated for by the use of adequate amounts of "good quality" protein (usually fortified with methionine) and by the addition of supple-ments (see Tables 1, 2, and 3). However, vitamin (especially thiamin) losses can be controlled to some extent by bringing the water content of the diet to about 25% before sterilization.[23] Autoclaving as described will destroy less than 10% of phytate in grain-containing diets.[24] Sterilization temper-atures higher than 121°C, if they can be achieved, will allow substantially reduced sterilization times and will generally reduce the loss of nutrients. It has been established that little loss of minerals due to leaching occurs during the steam phase of the sterilization process (Wostmann, B.S., un-published results).

Whenever sterilization by radiation is available, it should be regarded as the method of choice for solid diets.[25] Usually gamma radiation is used, and an accumulation of 4 Mrad is considered sufficient to guarantee sterility. At that dose, losses of protein quality appear to be very small, but limited losses occur in Vitamins A, E, B_1, B_6, and B_{12}. As in the case of steam sterilization, vitamin supplementation will ensure adequate amounts of vitamins to be available in the sterile diets. As mentioned before, some loss of phytate also occurs during this process.[23] Since irra-diation will be done away from the isolator facility, the diet usually is prepared in sealed plastic bags; these will have to be introduced into the isolators via surface sterilization procedures.[26]

It should be kept in mind that autoclaving, and to a lesser extent radiation sterilization, may add to the antigenicity of the diet. Also, casein has been found to be a notorious source of dead microorganisms, which again will add to dietary antigenicity.[27] Furthermore, diets containing substantial amounts of mono- and/or disaccharides cannot be successfully autoclaved because extensive chemical reaction between dietary components will take place, especially between carbohydrates and proteins. This is true to a much lesser extent for sterilization by radiation.

METABOLIC REQUIREMENTS OF GERMFREE RATS AND GERMFREE MICE — CONTRIBUTIONS OF THE MICROFLORA

It is, of course, impossible to clearly separate the above items because of the many effects of the presence of a microflora on the physiology of the gastrointestinal tract, which in turn will affect the utilization of dietary nutrients. However, a number of qualitative, and even some semiquantitative observations are possible.

Dietary Energy Requirements

As elaborated in Chapter IV and mentioned again in the beginning of this chapter, both GF rats and GF mice appear to have more efficient metabolic machinery than their CV counterparts as demonstrated by lower resting O_2 use, lower food intake by young growing animals, and various other indicators.

Proteins and Amino Acids

The available data point to lower qualitative and quantitative requirements of GF rats and GF mice. Dubos and Schaedler[28] showed early on that ex-GF mice associated with a controlled, nonpathogenic microflora could survive on a diet of corn, whereas mice harboring a "normal" microflora would lose weight and eventually die. Even when fed a diet containing casein in marginal quantities, the "clean" mice always grew much better than the CV "dirty" mice. Along the same lines, Gustafsson[29] reported acceptable growth of GF rats maintained on diets containing only 6 to 8% protein, an amount too low to sustain growth in CV rats. Refining the approach, Stoewsand et al.[30] fed more defined diets containing increasing amounts of lysine to GF mice, ex-GF mice associated with a nonpathogenic microflora originating in Dubos' laboratory,[31] and CV mice. Both the GF mice and the GN mice grew maximally on 2/3 of the lysine intake needed by the CV mice for optimal growth.

Reddy et al.[32] have reported a slightly higher nitrogen retention in young GF rats. Also, GF mice appear to absorb certain amino acids faster than CV mice,[33] although in view of the fact that most dietary protein is 95% utilized, it is doubtful whether the latter observation would have any effect on protein requirements.

There seems to be no reason to assume, once amino acids and small peptides have been absorbed, that any qualitative or quantitative difference exists in their metabolism. This may be illustrated by the fact that Woods and Goldman[34] demonstrated that GF rats can utilize the alpha-hydroxy analogues of phenylalanine and leucine as well as CV rats. To what extent the lower protein requirements, either qualitative or quantitative, are to be ascribed to the GF rat's more efficient metabolism (as described in Chapter IV) or to an absence of nutrient-microflora interaction, or to the absence of the stress of "physiological inflammation" in the gut,[35] remains unclear. Studies by Klasing et al.[36,37] suggest that at least in the chicken the latter may be a factor to be considered.

Vitamin B Complex — Microflora Production

In 1951 Lih and Baumann reported a B-vitamin-sparing effect of antibiotics.[38] They found that the addition of antibiotics, especially penicillin, to a diet limited in thiamin, riboflavin, or folic acid would substantially increase the growth of weanling rats. Weanling rats fed a diet deficient in thiamin, but otherwise complete, stopped growing after about 2 weeks. Weight loss followed, and death occurred within 4 weeks. Addition of penicillin to the deficient diet made limited growth possible, and resulted in a marginal increase in the thiamin content of the liver.[39] A few years later Wostmann et al. showed that the addition of penicillin to a diet containing only 0.4 mg/kg of thiamin had no effect in GF rats, but a positive effect in comparable CV rats fed the same sterilized diet[40] (Figure 1). This study found a substantial increase in liver thiamin concentration of the CV rats fed the penicillin-supplemented thiamin-deficient diet. The results clearly indicated a positive contribution of the microflora, although other effects of antibiotic treatment such as a more permeable gut wall, enlargement of the cecal mass, and increased coprophagia all may have played a role. Taken together, these studies clearly pointed to a potential of the microflora to produce B vitamins, although the nutritional importance of the phenomenon remained questionable.

Additional studies by Wostmann and Knight[41] showed that after administration of ^{35}S-labeled Na_2SO_4 via gastric intubation to CV rats maintained on a complete diet, ^{35}S-labeled thiamin could be detected in small but significant amounts in the lower small intestine and in larger amounts in the cecum. However, no ^{35}S-labeled thiamin could be detected in the

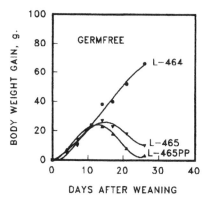
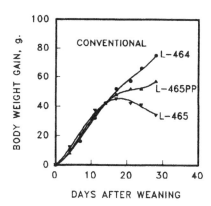

FIGURE 1
Body weight gain of male germfree and conventional Lobund-Wistar rats fed a complete diet (L-464) or a thiamin-deficient diet without (L-465) or with (L-465PP) added penicillin. See text.

liver. Considering the specific activity of the thiamin isolated from the cecum it was concluded that under these conditions, even though limited coprophagia might have taken place on the supposedly complete diet, the contribution of the microflora-synthesized thiamin to the host was low and relatively unimportant. Maintained on a thiamin-deficient diet, CV rats would have eaten more of their feces and microbial production of thiamin could have become a significant factor, as indicated by their somewhat longer survival and by the higher liver thiamin content of CV than of GF rats.[40] In later studies the authors estimated the microbial thiamin production in the CV rat and concluded that intestinal and cecal production, had it been available, would have covered most of the metabolic requirements of the animal.[42] Analysis of the thiamin present in the cecal contents showed it to be firmly bound as carboxylase, with only a small fraction in solution as free thiamin.

Eventually it became obvious that all the B vitamins could be synthesized by the intestinal microflora. In 1969 Coates and co-workers stated categorically that all B-complex vitamins can be synthesized by CV birds, but "the birds derived little benefit from the synthesized vitamins, with the possible exception of folic acid."[43] Earlier studies by Daft *et al.* had already shown that in the CV rat its microflora production could cover requirements of folic acid, and possibly some of pantothenic acid,[44] although Mameesh and co-workers had found that pantothenic acid was available only via coprophagia.[45] It appears that in CV rats the microflora also contributes to some extent the requirements for biotin[46] and vitamin B_{12},[47,48] in addition to providing all the folic acid needed for normal growth and development. However, both Coates *et al.*[43] and Ikeda *et al.*[49] express doubt as to the availability of the microflora-produced biotin. A limited contribution to the B_6 status of the host is indicated by the work of Sumi *et al.*[50] and of Ikeda *et al.*[51] This microflora contribution appears to be of the same nature as that of thiamin, largely dependent on the fact that the feeding of a deficient diet greatly increases coprophagia.

In those cases where in the CV animal the microflora would cover a substantial part of the requirements, the GF animal was used to determine its true metabolic requirements. Daft *et al.* estimated the pantothenic acid requirement of the GF rat to be in the order of 100 to 300 µg/day.[44] Valencia *et al.*[52] established that a minimum of 15 to 20 ng of B_{12} per day was needed to maintain reproduction of GF rats. No firm data are available for the biotin requirement, since it appears very difficult to obtain biotin-free solid diets. The amount added to our chemically defined diet (see Chapter VI) would work out to around 30 µg/day for the GF rat, although in view of the difficulty of producing biotin deficiency, this estimated amount may be much more than actually needed. A recent study with GF ICR mice by Komai *et al.*, using fetal biotin content and malformation of the fetuses as criteria, came to a requirement of the pregnant mother of 50 to 100 µg/100 g diet, or an estimated 2.8 µg biotin per day.[53]

In 1951 it was reported that the GF rat could synthesize choline,[54] but it remained questionable to what extent this could cover the animal's requirements. Available data suggest that the amount produced might be marginal,[55,56] making it advisable to add choline to the dietary vitamin supplement of the GF rodent.

In conclusion, it appears that with the exception of the instances mentioned above, the microflora production of the B vitamin group will add little to cover requirements to a diet otherwise supporting satisfactory growth and reproduction. This means that established requirement data for CV rats and CV mice will cover the needs of their GF counterparts, provided that losses occurring during diet sterilization are covered. The thiamin requirements of both the GF and CV Lobund-Wistar male rats were found to be around 18 µg/day.[42]

Fat-Soluble Vitamins

Physicians have known for a long time that vitamin K was required only for the newborn, and possibly in a case of prolonged treatment with antibiotics. It is actually impossible to produce a vitamin K deficiency in the CV rodent, even on a diet totally devoid of any form of vitamin K. Microflora production is always available to an extent that it appears to fully cover requirements. Only with the advent of the GF animal was it possible to establish systemic requirements and to judge the efficiency of the various forms of the vitamin. Studies with GF rats required extensive lipid extraction of dietary components to even produce a deficiency. Data obtained by both Gustafsson *et al.*[57] and Wostmann *et al.*[58] showed a requirement of approximately 3 µg vitamin K_1 per day. They also indicated a definite superiority of vitamin K_1 (phylloquinone) over vitamin K_3

(menadione). On a molar basis, requirements for the latter are about 10 times as great as for vitamin K_1.

Much more surprising was the observation by Bieri et al.[59] that GF rats would grow and survive for almost a year on diets containing hardly any vitamin A. Growth of the CV Sprague-Dawley rats, on the other hand, was arrested at about 200 g, and typical symptoms of Vitamin A deficiency were obvious. Under these conditions the CV rats died within 2 months.[60] The data suggest that the well-established role of vitamin A in tissue integrity and the related resistance to infection play a role here, since when these GF rats became accidentally contaminated they died within days. Recently Zhao and Ross[61] studied the extent to which vitamin A deficiency affects the immune system. They report that repletion with retenoic acid reestablishes the number of circulating lymphocytes and, in addition, may stimulate NK cell function.

To our knowledge, no differences between the vitamin D and vitamin E requirements of GF and CV rodents have been reported to date.

Minerals

Thus far, differences have been established in the way Ca, Mg, Zn, Fe, Cu, and P are handled by the GF animal, although other less obvious differences may be detected in the future. The significantly higher retention of Ca and Mg has been ascribed largely to the higher bile acid content of the GF gut, resulting in better absorption of Ca- and Mg-fatty acid complexes.[62] Increased levels of brush border Ca^{2+}- and Mg^{2+}-stimulated ATPases of the on the average older GF enterocyte could be an additional factor,[63] as could an increase in paracellular absorption[64] because of easier passage through the thinner lamina propria of the GF animal. Since Ca and Mg blood levels are quite rigorously controlled, this increased absorption then causes the GF rat to have a heavier skeleton, the necessary phosphate being made available by a secondary reduction in urinary excretion[65] (see Chapter III). In later diet formulations the amount of Ca and Mg in the mineral supplements was reduced to prevent abnormal calcification and calculi formation.

The presently available data on Zn metabolism suggest that the GF animal may again be in a more advantageous position. Smith et al.,[66] feeding a casein-starch diet containing only 3.7 ppm Zn to GF and CV Sprague-Dawley rats, report that the GF animals grew better and had a much higher Zn concentration in plasma and bone than CV rats fed the same Zn-deficient diet. Increasing the dietary Zn to 117 ppm eliminated differences in growth and plasma Zn. Reddy et al.[67] fed a similar casein-starch diet containing 96 ppm Zn to male Lobund-Wistar and found a slightly, but not significantly, better absorption and retention of Zn by the GF animals and a 24% higher retention of Zn in the skeleton. The latter, however, may have resulted from the higher Ca and Mg retention of the

GF rat fed this type of diet (see above) which by causing a heavier skeleton would have deposited additional Zn. Data obtained in the Lobund Aging Study (see Chapter I) using the natural ingredient diet L-485 (Table 2) whose Zn content was only 32 ppm, showed serum Zn concentrations in GF rats to be about 20% higher than in CV rats, although both levels (1.48 and 1.24 mg/l, respectively) were well within the range reported in the literature for the CV animal.[68] Diet L-485 supports excellent growth, reproduction, and life span at the above-mentioned Zn level (see Chapter II). According to Nishimoto and Padila,[69] this diet does not produce the significantly heavier skeleton with its potential Zn requirement seen in GF rats fed the earlier casein-starch diets. Combined with possibly lower systemic requirements, this could lead to a somewhat higher serum Zn concentration, especially in older GF rats. Taken together, the data suggest that the GF rat, for reasons not quite clear, actually has a lower metabolic requirement for Zn.

Thus, in the case of Ca, Mg, and Zn we see differences in absorption and internal homeostasis which could make small differences in dietary requirements, but present diet formulas for CV animals usually contain levels of these elements that will not cause deficiency conditions in their GF counterparts. However, Fe nutrition appears to present a different story. When GF rabbits were fed a diet supplemented with ferrous citrate $(FeC_6H_5O_7 \cdot 5H_2O)$ they would become severely anemic, while their CV counterparts and also GF and CV rats did well on that same diet.[70] The inclusion of soy bean meal in the diet, with its high content of organically bound Fe, solved the problem (see Chapter III). To further investigate this matter, Reddy et al.[71] studied organ distribution of heme iron and storage iron in GF and CV rats maintained on colony diet L-462 (Table 1), a diet based on grains and milk proteins that contained 800 ppm of Fe and 100 ppm Cu. This diet had been used in colony production for almost 10 years before being replaced by the much cheaper diet L-485 (Table 2). Hemoglobin and hematocrit values of the GF rats fed diet L-462 were comparable to those of the CV animals, but plasma Fe and Cu were slightly but significantly lower. Total body Fe concentration of the GF rat was around 75% of that of the CV rat, with the liver of the GF animal containing more Fe but spleen and kidney containing much less Fe. Within the organs, distribution of Fe among the heme Fe, ferritin Fe, and hemosiderin Fe was similar. Cu was distributed in a similar fashion as Fe. It would appear that in the GF rat the liver may be a major Fe storage organ.

Obviously, major differences in Fe homeostasis exist between GF and CV animals. Although not quite understood, they point to problems with Fe availability, presumably due at least in part to the more positive oxidation-reduction potential in the GF gut.[8,72] These data stress the need for an adequate Fe supply covered at least in part by organically bound Fe as supplied, e.g., by soy products. As mentioned earlier (Chapter III), the dietary source of Fe is much more important for the GF rabbit than for the GF rat, the latter being relatively insensitive to the form in which Fe is made available.[73] Since GF mice have been and are still being used extensively in immunological research, this matter is of special importance because inadequate Fe nutrition may impact on many facets of the immune potential, as mentioned in a recent study on the role of natural killer cells in the protection against cancer.[74]

Inositol, Ubiquinone, and Queuine

Inositol is produced by the body, but there remains a certain amount of doubt as to whether this production will cover requirements under all conditions.[75] Therefore it is generally included in dietary supplements for GF animals.

Ubiquinone (coenzyme Q) is a benzoquinone which functions in the electron-transport chain. At one time it was considered as a possible vitamin, because of its apparent potential to protect against vitamin E deficiency. A study by Bieri and McDaniel[76] showed that complete biosynthesis of ubiquinone occurs in the tissue of the rat, and that no exogenous source of the benzoquinone ring is needed.

The nucleoside queuosine is a modified derivative of guanosine, and is inserted posttranscriptionally as queuine in the "wobble position" of the anticodon of $tRNA^{Asp}$, $tRNA^{Asn}$, $tRNA^{His}$, and $tRNA^{Tyr}$. Its presence supposedly facilitates the proper translation of mRNA.[77] It is normally provided by microbial synthesis, but may have to be provided by the diet under GF conditions. Although data from the study of chemically defined diets indicate normal growth of GF mice in its absence (see Chapter VI), under conditions of increased stress, e.g., reproduction, this factor may improve performance of the GF animal.

REFERENCES

1. Nuttal, G. H. F. and Thierfelder, H., Thierisches Leben ohne Bakterien im Verdauungskanal, *Z. Physiol. Chem.*, 21, 109, 1895-1896.
2. Cohendy, M. and Wollman, E., Expériences sur la vie sans microbes, *C. R. Acad. Sci. Paris*, 26, 106, 1914.
3. Glimstedt, G., Das Leben ohne Bakterien. Sterile Aufziehung von Meerschweinchen, *Anat. Anz.*, 75, 79, 1932.
4. Reyniers, J. A., Trexler, P. C., and Ervin, R. F., Rearing germfree albino rats, *Lobund Report #1*, Notre Dame University Press, Notre Dame, IN, 1946.
5. Gustafsson, B. E., Germfree rearing of rats: general technique, *Acta Pathol. Microbiol. Scand.*, Suppl. 73, 1, 1948.
6. Pleasants, J. R., Rearing germfree Cesarian-born rats, mice, and rabbits through weaning, *Ann. N.Y. Acad. Sci.*, 78, 116, 1959.
7. Miyakawa, M., Studies of rearing germfree rats, in *Advances in Germfree Research and Gnotobiology*, Miyakawa, M. and Luckey, T. D., Eds., CRC Press, Boca Raton, FL, 1967, 48.

8. Phillips, B. P., Wolfe, P. A., and Gordon, H. A., Studies on rearing the guinea pig germfree, *Ann. N.Y. Acad. Sci.*, 78, 183, 1959.

9. Pleasants, J. R., Reddy, B. S., Zimmerman, D. R., Bruckner-Kardoss, E., and Wostmann, B. S., Growth, reproduction and morphology of naturally born, normally suckled germfree guinea pigs, *Z. Versuchstierk.*, 9, 195, 1969.

10. Pleasants, J. R., Johnson, M. H., and Wostmann, B. S., Adequacy of chemically defined, water-soluble diet for germfree BALB/c mice through successive generations and litters, *J. Nutr.*, 116, 1949, 1986.

11. Coates, M E., Ed., *ICLAS Guidelines on the Selection and Formulation of Diets for Animals in Biomedical Research,* Institute of Biology, London, 1987, 1.

12. Pleasants, J. R., Animal production and rearing. I. Small laboratory animals, in *The Germfree Animal in Research,* Coates, M. E., Ed., Academic Press, London, 1968, 48.

13. Bartizal, K. F., Beaver, M. H., and Wostmann, B. S., Cholesterol metabolism in gnotobiotic gerbils, *Lipids*, 17, 791, 1982.

14. Larner, J. and Gillespie, R. E., Gastrointestinal digestion of starch, *J. Biol. Chem.*, 225, 279, 1957.

15. Wostmann, B. S., Nutrition of the germfree mammal, *Ann. N.Y. Acad. Sci.*, 78, 175, 1959.

16. Gustafsson, B. E., Lightweight stainless steel systems for rearing germfree animals, *Ann. N.Y. Acad. Sci.*, 78, 17, 1959.

17. Kellogg, T. F. and Wostmann, B. S., Stock diet for colony production of germfree rats and mice, *Lab. Anim. Care*, 19, 812, 1969.

18. Garnier, H. and Sacquet, E., Absorption apparente et rétention du sodium, du potassium, du calcium et du phosphore chez le rat axénique et chez le rat holoxénique, *C. R. Acad. Sci. Paris*, 269, 379, 1969.

19. Wostmann, B. S. and Kellogg, T. F., Purified starch-casein diet for nutritional research with germfree rats, *Lab. Anim. Care*, 17, 589, 1967.

20. Wostmann, B. S., Bruckner-Kardoss, E., Beaver, M., Chang, L., and Madsen, D., Effects of dietary lactose at levels comparable to human consumption on cholesterol and bile acid metabolism of conventional and germfree rats, *J. Nutr.*, 106, 1782, 1976.

21. Iwasaki, K., Gleiser, C. A., Masoro, E. J., McMahan, C. A., Seo, E., and Yu, B. P., The influence of dietary protein source on longevity and age-related disease processes of Fischer rats, *J. Gerontol.*, 43, B52, 1988.

22. Coates, M. E., Diets for germfree animals. I. Sterilization of diets, in *The Germfree Animal in Biomedical Research,* Coates, M. E. and Gustafsson, B. E., Eds., Laboratory Animal Handbook 9, Laboratory Animals Ltd., London, 1984, 85.

23. Zimmerman, D. R. and Wostmann, B. S., Vitamin stability in diets sterilized for germfree animals, *J. Nutr.*, 79, 318, 1963.

24. Yoshida, T., Shinoda, S. and Watarai, S., Destruction of phytic acid in diet for germfree rats and mice during autoclaving and gamma irradiating, *Jpn. J. Nutr.*, 38, 215, 1980.

25. Ley, F. J., Bleby, J., Coates, M. E., and Paterson, J. S., Sterilization of laboratory animal diets using gamma radiation, *Lab. Anim.*, 3, 221, 1969.

26. Trexler, P. C., Isolator technology for microbic contamination control, *Z. Versuchstierk.*, 10, 121, 1968.

27. Wostmann, B. S., Pleasants, J. R., and Bealmear, P., Dietary stimulation of immune mechanisms, *Fed. Proc.*, 20, 1779, 1971.

28. Dubos, R. J. and Schaedler, R. W., The effect of the intestinal flora on the growth rate of mice and their susceptibility to experimental infection, *J. Exp. Med.*, 111, 407, 1960.

29. Gustafsson, B. E., Introduction of specific microorganisms into germfree animals, in *Nutrition and Infection,* Ciba Study Group No. 31., Wolstenholme, G. E. W. and O'Connor, M., Eds., Little, Brown, Boston, 1967, 16.

30. Stoewsand, G. S., Dynsza, H. A., Ament, D., and Trexler, P. C., Lysine requirements of the growing gnotobiotic mouse, *Life Sci.*, 7, 689, 1968.

31. Schaedler, R. W., Dubos, R., and Costello, R., Association of germfree mice with bacteria isolated from normal mice, *J. Exp. Med.*, 122, 77, 1965.

32. Reddy, B. S., Wostmann, B. S., and Pleasants, J. R., Protein metabolism in germfree rats fed chemically defined, water-soluble and semi-synthetic diet, in *Germfree Biology: Experimental and Clinical Aspects*, Mirand, E. and Back, X., Eds., Plenum Press, New York, 1969, 301.

33. Herskovic, T., Katz, J., Floch, M. H., Spencer, R. D., and Spiro, H. M., Small intestinal absorption and morphology in germfree, monocontaminated and conventionalized mice, *Gastroenterology*, 52, 1136, 1967 (Abstr.).

34. Woods, M. N. and Goldman, P., Replacement of L-phenylalanine and L-leucine by alpha-hydroxy analogues in the diet of germfree rats, *J. Nutr.*, 109, 738, 1979.

35. Gordon, H. A., Bruckner-Kardoss, E., Staley, T. E., Wagner, M., and Wostmann, B. S., Characteristics of the germfree rat, *Acta Anat.*, 64, 367, 1966.

36. Klasing, K. C. and Austic, R. E., Changes in protein degradation in chickens due to an inflammatory challenge, *Proc. Soc. Exp. Biol. Med.*, 176, 292, 1984.

37. Klasing, K. C., Laurin, D. E., Peng, R. K., and Fry, D. M., Immunologically mediated growth depression in chicks: influence of feed intake, corticosterone and interleukin-1, *J. Nutr.*, 117, 1629, 1987.

38. Lih, H. and Baumann, C. A., Effects of certain antibiotics on the growth of rats given limiting B vitamins by mouth or by injection, *J. Nutr.*, 45, 143, 1951.

39. Jones, J. D. and Baumann, C. A., Relative effectiveness of antibiotics in rats given limiting B vitamins by mouth or by injection, *J. Nutr.*, 57, 61, 1955.

40. Wostmann, B. S., Knight, P. L., and Reyniers, J. A., The influence of orally administered penicillin upon growth and liver thiamine of growing germfree and normal stock rats fed a thiamine-deficient diet, *J. Nutr.*, 66, 577, 1958.

41. Wostmann, B. S. and Knight, P. L., Synthesis of thiamine in the digestive tract of the rat, *J. Nutr.*, 74, 103, 1961.

42. Wostmann, B. S., Knight, P. L., and Kan, D. F., Thiamine in germfree and conventional animals. Effect of the intestinal microflora on thiamine metabolism of the rat, *Ann. N.Y. Acad. Sci.*, 98, 516, 1962.

43. Coates, M. E., Ford, J. E., and Harrison, G. F., Intestinal synthesis of vitamins of the B complex in chicks, *Br. J. Nutr.*, 22, 493, 1968.

44. Daft, F. S., McDaniel, E. G., Harman, L. G., Romine, L. K., and Hegner, J. L., Role of coprophagia in utilization of B-vitamins synthesized by intestinal bacteria, *Fed. Proc.*, 22, 129, 1963.

45. Mameesh, M. S., Webb, R. E., Norton, H. W., and Connor Johnson, B., The role of coprophagy in the availability of vitamins synthesized in the intestinal tract with antibiotic feeding, *J. Nutr.*, 69, 81, 1959.

46. Cherruau, B., Sacquet, E., Mangeot, M., Demelier, J. F., and Lemonnier, A., Carence en biotine chez le rat axénique et acidémique propionique, *Ann. Nutr. Metab.*, 27, 48, 1983.

47. Valencia, R. and Sacquet, E., La carence en vitamine B_{12} chez l'animal amicrobien (anomalies tératogènes et signes hématologiques), *Ann. Nutr. Aliment.*, 22, 71, 1968.

48. Oace, S. M. and Abbott, J. M., Methylmalonate, formiminoglutamate and aminoimidazolecarboxamide excretion of vitamin B_{12}-deficient germfree and conventional rats, *J. Nutr.*, 102, 17, 1972.

49. Ikeda, M., Iwai, M., Sato, H., and Sakakibara, B., The role of intestinal flora in biotin deficiency in conventional and germfree mice fed a purified biotin-deficient diet without supplementation with egg white, *J. Clin. Biochem. Nutr.*, 17, 103, 1994.

50. Sumi, Y., Miyakawa, M., Kanzaki, M., and Kotake, Y., Vitamin B-6 deficiency in germfree rats, *J. Nutr.*, 107, 1707, 1977.

51. Ikeda, M., Hosotani, T., Kurimoto, K., Mori, T., Ueda, T., Kotake, Y., and Sakakibara, B., Differences of the metabolism related to the B_6-dependent enzymes among vitamin B_6-deficient germfree and conventional rats, *J. Nutr. Sci. Vitaminol.*, 28, 131, 1979.

52. Valencia, R., Sacquet, E., and Jaquot, R., Les charactères de l'avitaminose B_{12} chez le rat axénique et le rat normal, *J. Physiol. Paris*, 60 (Suppl. 2), 561, 1968.

53. Komai, M., Fukasawa, H., Furukawa, Y., and Kimura, S., Metabolic characteristics of primary biotin deficiency established in germfree mice, *Microecol. Ther.*, 20, 63, 1990.

54. du Vigneau, V., Ressler, C., Rachele, J. R., Reyniers, J. A., and Luckey, T. D., The synthesis of "biologically labeled" methyl groups in the germfree rat, *J. Nutr.*, 45, 361, 1951.

55. Levenson, S. M., Nagler, A. L., Geever, E. F., and Seifter, E., Acute choline deficiency in germfree, conventional and open-animal-room rats. Effects of neomycin, chlortetracycline, vitamin B_{12} and coprophagia prevention, *J. Nutr.*, 95, 247, 1968.

56. Kwong, E., Fiala, G., Barnes, R. H., Kan, D., and Levenson, S. M., Choline biosynthesis in germfree rats, *J. Nutr.*, 96, 10, 1968.

57. Gustafsson, B. E., Daft, F. S., McDaniel, E. G., Smith, J. C., and Fitzgerald, R. J., Effects of vitamin K-active compounds and intestinal microorganisms in vitamin K-deficient rats, *J. Nutr.*, 78, 461, 1962.

58. Wostmann, B. S., Knight, P. L., Keely, L. L., and Kan, D. F., Metabolism and function of thiamine and naphthoquinones in germfree and conventional rats, *Fed. Proc.*, 22, 120, 1963.

59. Bieri, J. G., McDaniel, E. G., and Rogers, W. E., Survival of germfree rats without vitamin A, *Science*, 163, 574, 1969.

60. Rogers, W. E., Bieri, J. G., and McDaniel, E. G., Vitamin A deficiency in the germfree state, *Fed. Proc.*, 30, 1773, 1971.

61. Zhao, Z. and Ross, A. C., Retenoic acid repletion restores the number of leukocytes and their subsets and stimulates natural cytotoxicity in vitamin A-deficient rats, *J. Nutr.*, 125, 2064, 1995.

62. Desmarne, Y., Flanzy, J., and Sacquet, E., The influence of gastrointestinal flora on digestive utilization of fatty acids in rats, in *Germfree Research: Biological Effects of Gnotobiotic Environments. Proceedings of the IVth International Symposium on Germfree Research,* Heneghan, J. B., Ed., Academic Press, New York, 1973, 553.

63. Reddy, B. S., Studies on the mechanism of calcium and magnesium absorbtion in germfree rats, *Arch. Biochem. Biophys.*, 149, 15, 1972.

64. Stein, W. D., Facilitated diffusion of calcium across rat intestinal epithelial cell, *J. Nutr.*, 122, 651, 1992.

65. Reddy, B. S., Pleasants, J. R., and Wostmann, B. S., Effect of intestinal microflora on calcium, phosphorus and magnesium metabolism in rats, *J. Nutr.*, 99, 353, 1969.

66. Smith, J. C., McDaniel, E. G., McBean, L. D., Doft, F. S., and Halsted, J. A., Effect of microorganisms upon zinc metabolism using germfree and conventional rats, *J. Nutr.*, 102, 711, 1972.

67. Reddy, B. S., Pleasants, J. R., and Wostmann, B. S., Effect of intestinal microflora on iron and zinc metabolism, and on activities of metalloenzymes in rats, *J. Nutr.*, 102, 101, 1972.

68. Wostmann, B. S., Wong, F. R., and Snyder, D. L., Serum zinc in aging germfree and conventional rats, *Proc. Soc. Exp. Biol. Med.*, 199, 218, 1992.

69. Nishimoto, S. K. and Padila, S. M., Age-related changes in organic and inorganic bone matrix constituents. The effect of environment and diet, in *Dietary Restriction and Aging*, Snyder, D. L., Ed., *Prog. Clin. Biol. Res.*, 287, 87, 1989.

70. Reddy, B. S., Pleasants, J. R., Zimmerman, D. R., and Wostmann, B. S., Iron and copper utilization in rabbits as affected by diet and germfree status, *J. Nutr.*, 87, 189, 1965.

71. Reddy, B. S., Wostmann, B. S., and Pleasants, J. R., Iron, copper and manganese in germfree and conventional rats, *J. Nutr.*, 86, 159, 1965.

72. Wostmann, B. S. and Bruckner-Kardoss, E., Oxydation-reduction potentials in cecal contents of germfree and conventional rats, *Proc. Soc. Exp. Biol. Med.*, 121, 1111, 1966.

73. Reddy, M. B. and Cook, J. D., Absorption of nonheme iron in ascorbic acid-deficient rats, *J. Nutr.*, 124, 882, 1994.

74. Spear, A. T. and Rothman Sherman A., Iron deficiency alters DMBA-induced tumor burden and natural killer cell cytotoxicity in rats, *J. Nutr.*, 122, 46, 1992.

75. Freinkel, N. and Dawson, R. M. C., The synthesis of *meso*inositol in germfree rats and mice, *Biochem. J.*, 81, 250, 1961.
76. Bieri, J. G. and McDaniel, E. G., Ubiquinone (coenzyme Q) in the germfree rat, *Biochim. Biophys. Acta*, 56, 602, 1962.
77. Reyniers, J. P., Pleasants, J. R., Wostmann, B. S., Katze, J. R., and Farkas, W. R., Administration of exogenous queuine is essential for the biosynthesis of the queuosine-containing transfer RNAs in the mouse, *J. Biol. Chem.*, 256, 11591, 1981.

Chapter VI

THE CHEMICALLY
DEFINED DIET*

GENERAL ASPECTS

Although both GF rat and GF mouse colonies had been established via hand-feeding Cesarian-born young when the author came to the Lobund Laboratory in 1955, hand-feeding was still considered more of an art than a science. It was a very essential technique, but its results at the time, especially in mice, were far from satisfactory.[1] Weight gain on the cows' milk-based formulas was much below normal in GF rats, and it proved impossible to successfully reproduce the early results and hand-feed newborn mice through weaning (see Chapter I). Whereas normal rat milk contains 12% protein, the protein concentration in our hand-feeding formulas could not be raised over 6% without encountering severe digestive problems, as indicated by substantial amounts of undigested protein throughout the entire length of the small intestine.[2] Dr. Bengt Gustafsson at the University of Lund in Sweden has informed this author that he had encountered the same problem, with approximately the same limit for the protein concentration that could be incorporated in the milk formulas (personal discussion). As experiments with baby rats subsequently showed, the stress of hand-feeding resulted in premature secretion of HCl in the stomach. Normally, for the first two weeks of life the animal would depend on regurgitated intestinal enzymes, and only thereafter start to rely on stomach acid secretion and peptic digestion.[3] Hand-feeding caused premature acidification of stomach contents, resulting in pH values of 4 to 5,[2] a range in which neither the normally regurgitated proteolytic enzymes nor the prematurely secreted pepsin would function adequately. Obviously the problem was inherent to the hand-feeding procedure, and could

---

* Abbreviations used in this chapter, other than the usual GF and CV, are CD for chemically defined diet, NI for natural ingredient diet (in our own studies we always used diet L-485, see text); GF-CD, GF-NI, and CV-NI indicate GF and CV animals maintained on chemically defined, respectively, natural ingredient diets.

not be solved within the framework of that technique unless other ways were found to increase the availability of the necessary protein constituents.

In 1957 Greenstein *et al.*,[4] reported the development of a water-soluble, chemically defined diet (CD diet) for CV rats. Consultation with the authors led to our adoption of soluble formulas in which amino acids substituted for the apparently difficult to digest sterilized milk protein. These diets, based on glucose, amino acids, vitamins, and minerals could be sterilized by filtration, thus avoiding the chemical changes caused by other modes of sterilization. They needed only a supplement of defined fats and fat-soluble vitamins which could be filter-sterilized and fed separately.

Thus, the development of the CD diets started because of practical necessity, especially for the production of GF mice. The derivation of new GF mouse strains by hand-feeding was considered of special importance, because it would prevent the vertical transmission by cross-suckling of viruses indigenous to many strains of mice, including those already obtained germfree. It was also clear, however, that hand-feeding of defined diets could serve to refine our knowledge of nutritional requirements at every stage of the life cycle, and could give further insight into the role played by maternal milk in physiological and immunological maturation.

After considerable adjustment of the original Greenstein formula to meet both infant and GF conditions, it proved possible to hand-rear GF rats on a CD diet from the first day of life, but only if they had been nursed by GF rat mothers for at least a few hours after birth.[5] GF mice required days of maternal suckling before they could be successfully hand-reared on a CD diet. For the infant rodents, the predigested character of the diet was found to cause more problems than it solved. Its low molecular weight components forced a rather uncompromising choice between high osmolarity and high dilution of the diet mixture.[5]

The hand-feeding studies had, however, contributed to the rapid adaptation of the Greenstein formula to the needs of GF rodents, since infant animals respond quickly to dietary inadequacies. Meanwhile, developments in the biomedical sciences made it increasingly clear that colony-reared (mother-suckled) GF rodents weaned to CD diets could contribute much-needed definition to studies in nutrition, metabolism, physiology and, especially, immunology. Whereas GF animals fed the natural ingredient (NI) diets had lower levels of IgG and IgA globulins than CV animals,[6] IgG never reached the low levels anticipated. The possibility that the GF-CD mouse could make basic contributions to immunology became evident after it was found that ultrafiltration of the CD diet to remove components with a molecular weight above 10 kDa virtually eliminated antigenic stimulation. Despite the fact that these GF mice had been exposed early on to the high molecular weight components of maternal milk, their immune systems showed very little activation[7] and were expected to show even less stimulation if it became possible to raise subsequent generations entirely on the CD diet. This would open the way for both basic and applied immunological studies based on a truly primary immune response.

Research with GF-CD diets therefore concentrated on the development of chemically defined, water-soluble, antigen-free diets which could then sustain GF rodents through repeated litters and generations. This would prove the adequacy of the diet to produce a healthy, well-defined experimental animal, never exposed to anything undefined except its mother's milk. Because of the high cost of the diet, most of this development phase was carried out with CFW and later with an inbred strain of C3H mice. A few of the earlier experiments with GF-CD rats indicated they would respond similarly to mice.

DEVELOPMENT OF THE CHEMICALLY DEFINED DIET

GF status itself dictated the early changes from the formulation of Greenstein et al.[4] Hemorrhages in the first GF-CD rats confirmed what had been found earlier in GF rats fed semisynthetic diets — compared to CV rats, GF rats utilize menadione (vitamin K_3) very poorly.[8] Menadione had to be replaced by a natural K factor, phylloquinone (vitamin K_1). The synthetic emulsifier, Tween 80®, used by Greenstein et al.[9] to incorporate the lipids into their one-solution diet, caused more severe diarrhea in our GF animals than had occurred in their CV rats. Popliteal lymph nodes hypertrophied, and gamma globulin levels rose above CV levels.[10] Others later reported the antigenicity[11] and hepatotoxicity[12] of Tween 80®. To eliminate Tween 80®, the lipids were filtered and fed separately from the water solution of the nutrients.[13]

Greenstein et al. also used the synthetic compound cysteine ethyl ester in place of the poorly soluble cysteine.[4] In our work this ethyl ester changed the odor, taste, and color of the CD diet. It was deleted since animals can convert methionine to cysteine. Later, Levenson et al. showed that cysteine ethyl ester in a Greenstein-type diet caused pancreatic acinar cell atrophy to some extent in GF rats, and to a much greater extent in CV rats. In CV rats it also caused hemolytic anemia.[14] The above indicates the important role of the animal's microflora in determining its response to unnatural compounds.

In the same time frame of this development, the trace minerals Se, Cr, V, F, Sn, and Ni were reported to be essential. These were added to the diet in the recommended amounts. When linolenate was shown to be an essential fatty acid, our purified lipid mix modeled on corn oil was replaced by one modeled on soy oil[13] (Table 1). A syndrome of sudden death with some soft tissue calcification indicated a higher need for Mg than is usually recommended for CV animals.[10] This may be related to calcifications seen in GF C3H mice fed semipurified diets[15] and to mineral imbalances found in GF guinea pigs and rabbits.[16,17]

TABLE 1

Composition of Chemically Defined Diet L-489E14SE Per 100 g
Water-Soluble Solids in 300 ml Ultrapure H_2O

L-Leucine	1.90 g	Ca glycerophosphate	5.22 g
L-Phenylalanine	0.74 g	$CaCl_2 \cdot 2H_2O$	0.185 g
L-Isoleucine	1.08 g	Mg glycerophosphate	1.43 g
L-Methionine	1.06 g	K acetate	1.85 g
L-Tryptophan	0.37 g	NaCl	86.00 mg
L-Valine	1.23 g	Mn $(acetate)_2 \cdot 4H_2O$	55.40 mg
Glycine	0.30 g	Ferrous gluconate	50.00 mg
L-Proline	1.48 g	$ZnSO_4 \cdot H_2O$	40.60 mg
L-Serine	1.33 g	Cu $(acetate)_2 \cdot H_2O$	3.70 mg
L-Asparagine	1.03 g	Cr $(acetate)_3 \cdot H_2O$	2.50 mg
L-Arginine HCl	0.81 g	NaF	2.10 mg
L-Threonine	0.74 g	KI	0.68 mg
L-Lysine HCl	1.77 g	$NiCl_2 \cdot 3H_2O$	0.37 mg
L-Histidine HCl·H_2O	0.74 g	$SnCl_2 \cdot 2H_2O$	0.31 mg
L-Alanine	0.59 g	$(NH_4)_6Mo_7O_{24} \cdot 4H_2O$	0.37 mg
Na L-glutamate	3.40 g	Na_3VO_4	0.22 mg
Ethyl L-tyrosinate HCl	0.62 g	Co $(acetate)_2 \cdot 4H_2O$	0.11 mg
α-D-Dextrose	71.40 g	Na_2SeO_3	0.096 mg

B Vitamins (in mg)

Thiamine HCl, 1.23; pyridoxine HCl, 1.54; biotin, 0.25; folic acid, 0.37;
vitamin B_{12} (pure), 1.44; riboflavin, 1.85; niacinamide, 9.2; i-inositol, 61.6;
Ca pantothenate, 12.3; choline HCl, 310.

Amounts of Lipid Nutrients in One Measured Daily Adult Dose of 0.25 ml

Purified soy triglycerides, 0.22 g; retinyl palmitate, 4.3 μg (7.8 IU);
cholecalciferol, 0.0192 μg (0.77 IU); 2-ambo-α-tocopherol, 2.2 mg; 2-
ambo-α-tocopheryl acetate, 4.4 mg; phylloquinone, 48.0 μg. The fatty
acid content is 12% palmitate, 2% stearate, 24% oleate, 55% linoleate,
8% linolenate.

Because mother mice needed nesting material for the successful nursing of their young, filter paper strips had to be provided, albeit reluctantly. Hashimoto *et al.*[18] reported a similar need. We found that GF-CD mice consumed this material to the extent of 8% of their ingested solids. This addition to the diet reduced cecal enlargement and the incidence of cecal volvulus and trichobezoars.[19]

The levels of total and individual amino acids were modified according to literature reports of more effective patterns. The best and final pattern, however, was arrived at by comparing the plasma amino acid pattern of GF-CD diet mice with that of CV mice fed natural ingredient (NI) diet L-485.[20] Changes in the dietary amino acids were then made until the patterns matched.[19] The resulting formulation, which also included minor changes in inorganic components (see Table 1 in Pleasants *et al.*[13]), supported reproduction of C3H mice into the fifth generation, thereby indicating that it contained all nutrients essential for the survival of the species. It thus met Schultze's 1957 stipulation for proving nutritional adequacy: "a defined diet fed to a germfree animal."[21]

However, although these results were promising only the first and second litters in any single generation of these C3H mice contained the normal number of young and were adequately nursed. There was a rapid drop in numbers born alive and in percent weaned per litter starting with the third litter in any generation. Since this syndrome resembled a Se deficiency described earlier in the literature, Se content of the diet formula was increased fourfold, but the same pattern persisted.[22]

At this time, available support shifted to the BALB/c mouse strain, which has special value for the study of the immune response and for the production of monoclonal antibody. Reproductive data thus became available from a GF BALB/c mouse colony that had been changed to the latest CD diet formula, including the higher Se levels. In contrast to C3H mice, BALB/c mice showed no decline of reproductive performance in successive litters.[13] Since that time the BALB/c strain has been maintained on this latest CD formulation, with only minor modifications. Formula L-489E14SE is given in Table 1.

Procedures

The GF-CD mice are housed in vinyl Trexler-type isolators. The polycarbonate shoe-box-type mouse cages (Figure 1) have lids modified to hold the inverted diet and water bottles. The cages' plastic bottoms have been replaced with stainless steel wire mesh false bottoms above removable drip pans.

The water-soluble portion of the diet shown in Table 1 is ultrafiltered through an Amicon PM10 membrane in an Amicon TC3E thin-channel filter holder. Although the ultrafiltered diet is already free of bacteria, molds, viruses, and molecules larger than 10 kDa, it is passed into the isolator through a sterile nylon filter membrane of 0.2 μm pore size because it is the safest route through the isolator barrier. The diet is fed *ad libitum* in inverted brown bottles having 0.16-cm holes drilled in their lids (Figure 1). Ultrafiltered water is also available. Whatman Ashless Filter Paper # 41 provides ingestible fiber and also serves as bedding and nesting material. It is autoclaved for 25 min at 121°C or irradiated at 4.5 Mrad.

Purified soy-derived triglycerides provide the essential fatty acids plus readily available calories. To this end, soy oil has been converted to methylesters which are vacuum-distilled over a specific range, and converted back to triglycerides. Purified fat-soluble vitamins A, D, E, and K are added (Table 1). The mixture is not ultrafiltered on the assumption that vacuum distillation has achieved the same result. It is then passed into the isolator as described above for the water-soluble portion of the diet, and dispensed to the mice as single measured daily doses in stainless steel planchets welded to the stainless steel cage dividers (Figure 1).

FIGURE 1
Cage for germfree mice maintained on water-soluble, chemically defined diet showing collars for inverted bottles containing water (translucent) and diet (dark). Also seen are the small dishes for the lipid supplement, and the filter-paper bedding. See text. For further details see Wostmann and Pleasants.[23]

Further technical details involved in colony production of GF mice maintained on CD diets may be found in the publications by Pleasants et al.,[13] Hashimoto et al.,[18] and Wostmann and Pleasants.[23]

Present Status

In 1992 the GF-CD BALB/c colony was in its tenth generation. Growth is comparable to that seen with natural ingredient diet L-485.[20] There is a consistent 20% reduction in litter size compared with CV animals fed diet L-485 — so consistent that it may point to a physiological factor rather than a mild dietary deficiency. The intake of the CD diet during pregnancy might be physiologically limited by plasma amino acid effects, osmotic effects, and the combined pressure of an enlarged cecum and an enlarged uterus, although BALB/c females are able to double their intake between onset and end of pregnancy.[13] The absence of queuosine, normally produced by the intestinal microflora, may also have played a role here (see Chapter V).

The aforementioned use of filter paper as bedding material leads to the ingestion of a limited amount of this material, but resulted in reduction of the size of the cecum. The fibrous matter appears in a totally unaltered state in the feces. Death due to cecal volvulus, although infrequent, still does occur. Since most of the GF-CD BALB/c mice have been used in immunological studies and are at present used for the production of monoclonal antibodies, insufficient data on life span are available although a number of animals have lived beyond 18 months.

When resting oxygen consumption and heart size were determined earlier in GF-CD C3H mice, values were 38 and 27% higher, respectively, than in GF mice maintained on solid diet L-485.[19] This suggests that GF mice fed an amino acid diet require additional energy to utilize this major diet component. Other investigators have had similar results with humans[24] and rats.[25]

Although the quantity of filter paper ingested produced a certain amount of bulk in the gastrointestinal tract, a diet consisting almost exclusively of dextrose and amino acids must have an effect on the physiology of the gut. Yamanaka et al.[26] report that even when casein is replaced by a free amino acid mixture in a solid diet, the length of the small intestine is significantly reduced in both GF and CV mice. Morin et al.[27] fed young CV Sprague-Dawley rats a liquid amino acid-dextrose formula for only 8 days and saw a certain level of atrophy, measured as loss of functional tissue mass per centimeter of the intestinal tract, in comparison with chow-fed animals. This could explain earlier results by Reddy et al.,[28] who found a 25 to 30% increase in Ca absorption and retention when either GF or CV rats were fed one of the earlier CD formulas, presumably because of easier paracellular absorption. Thus, available data suggest that as far as nutrient uptake is concerned, the GF animal fed a CD diet may be comparable with, and possibly somewhat superior to, the animal fed a solid diet.

DIETARY ANTIGENICITY AND IMMUNE POTENTIAL

The early phases of the development of the CD diets had been carried out largely with our non-inbred GF CFW mice. Whereas the composition of the diet may not have been quite ideal, sufficient numbers of animals were produced to evaluate the diet in terms of one of our goals: the absence of dietary antigenicity. Immunoelectrophoresis of the serum of 70-day-old GF-CD mice showed virtual absence of IgG antibody, while comparable GF mice maintained on colony diet L-485 showed amounts not much less than found in their CV counterparts. Both granulocytes and mononuclear cell counts were 20 to 50% those of the GF mice fed the colony diet. However, after 1 year on the CD diet, GF CFW mice showed appreciable amounts of IgG.[29,30]

Later studies were therefore carried out with the inbred C3H/HeCr strain. In addition, filter-sterilization procedures were improved by the addition of aforementioned prefiltration through an Amicon Diaflo TC3 ultrafilter,[13] which removed all material over 10 kDa. This procedure, together with the elimination of possible animal-to-animal antigenicity, now resulted in the absence of detectable IgA and IgG levels by immunoelectrophoresis, even in 14-month-old C3H mice. It decreased the already low white blood cell counts even further. Only IgM values remained in the range of those found in GF and CV C3H mice maintained on colony diet L-485.[29]

In a comparable study, Hashimoto et al.[18] fed a similar CD diet formula to GF ICR mice and used antisera specific for the various immunoglobulins to quantitate their levels in 5- to 6-week-old animals. Again, IgM levels were only slightly decreased in the absence of dietary antigenic exposure. IgA was not detected in either the serum or intestinal wall of any of the GF mice, whatever their diet; it was detected only in CV mice. The serum IgG level of the GF-CD mice was one-tenth that of GF mice maintained on a solid diet, and one-hundredth that of CV mice.

However, after immunization using sheep red blood cells (SRBC), GF-CD C3H mice at the Lobund Laboratory showed an immune potential as least as good as that of GF-NI or CV-NI mice. This was expressed in their postinoculation levels of anti-SRBC plaque-forming cells, serum IgG, and hemagglutinin.[29] Following in vitro challenge with phytohemagglutinin and concanavalin A, the spleen cell response of GF-CD mice was at least equal to, and possibly slightly above that of GF and CV mice fed the NI diet.[31] In the nonchallenged state, GF-CD spleens showed an amount of IgM-containing cells equal to the CV-NI spleen, but only 20% as many IgG-containing cells and no IgA-containing cells.

At this point in time we had come to the conclusion that the CD diet L-489E14SE was at least qualitatively adequate for GF C3H mice. Body weights and appearance were satisfactory and the diet supported reproduction into the fifth generation, although as mentioned earlier, reproduction dropped sharply after the first two litters in any generation.[19] However, enough animals could be produced to meet the needs of more sophisticated immunological studies.

To this end we entered into collaboration with the Department of Cell Biology, Immunology and Genetics of Erasmus University, Rotterdam, The Netherlands. The resulting studies again showed that the immune potential of the GF-CD mouse was essentially unimpaired. The antibody-specificity repertoire of the background IgM-secreting cells was found comparable to that of CV C3H mice, suggesting that the generation of the IgM antibody repertoire was independent of external antigenic stimulation.[32] LPS-activated cultures of spleen and bone marrow cells of GF-CD

C3H mice produced similar, or possibly slightly higher, numbers of IgM- and IgG_1-producing clones than the cells of their CV counterparts fed NI diet L-485. Again, the specificity repertoire of the stimulated cells appeared not to be influenced by prior antigenic exposure of the intact animals.[33] The study also suggested that the capability of IgM-secreting cells to switch to secretion of IgG_1 was not affected by prior antigen exposure,[34] although this was not confirmed by later studies (see Chapter VII).

As mentioned earlier, continued interest in immunological studies led to a change to inbred BALB/c mice (BALB/cAnN, obtained from the germfree operation at the University of Wisconsin). Determination of "background" Ig-secreting cells and of the IgM antibody repertoire confirmed results obtained with C3H mice.[35] This BALB/c colony has provided a healthy, nutritionally defined animal model exposed only to an absolute minimum of exogenous stimulation. Therefore, with CD diet L489E14SE as given in Table 1, the major developmental phase of the CD diets came to an end.

Further details of the immunological studies done with CD mice are given in Chapter VII. Their use for the production of monoclonal antibodies is highlighted in Chapter XII.

REFERENCES

1. Pleasants, J. R., Rearing germfree Cesarian-born rats, mice and rabbits through weaning, *Ann. N.Y. Acad. Sci.*, 78, 116, 1959.
2. Wostmann, B. S., Nutrition of the germfree mammal, *Ann. N.Y. Acad. Sci.*, 77, 175, 1959.
3. Platt, B. B., *Proc. Nutr. Soc.*, 13, XVI, 1954.
4. Greenstein, J. P., Birnbaum, S. M., Winitz, M., and Otey, M. C., Quantitative nutritional studies with water-soluble, chemically defined diets. I. Growth, reproduction and lactation in rats, *Arch. Biochem. Biophys.*, 72, 396, 1957.
5. Pleasants J. R., Reddy B. S., and Wostmann B. S., Improved hand-rearing methods for small rodents, *Lab. Anim. Care*, 14, 37, 1964.
6. Bealmear, P. M., Holterman, O. A., and Mirand, E. A., Influence of the microflora on the immune response, in *The Germfree Animal in Biomedical Research*, Coates, M. E. and Gustafsson, B. E., Eds., Laboratory Animal Handbooks 9, Laboratory Animals Ltd., London, 1984, 347.
7. Benner, R., van Oudenaren, A., Haaijman, J. J., Slingerland-Teunissen, J., Wostmann, B. S., and Hijmans, W., Regulation of the "spontaneous" (background) immunoglobulin synthesis, *Int. Arch. Allerg. Appl. Immunol.*, 66, 404, 1981.
8. Wostmann, B. S., Knight, P. L., Keeley, L. L., and Kan, D. F., Metabolism and function of thiamine and naphthoquinones in germfree and conventional rats, *Fed. Proc.*, 22, 120, 1963.
9. Greenstein, J. P., Otey, M. C., Birnbaum, S. M., and Winitz, M., Quantitative nutritional studies with water-soluble, chemically defined diets. X. Formulation of a nutritionally complete liquid diet, *J. Natl. Cancer Inst.*, 24, 211, 1960.
10. Pleasants, J. R., Reddy, B. S., and Wostmann, B. S., Qualitative adequacy of a chemically defined liquid diet for reproducing germfree mice, *J. Nutr.*, 100, 498, 1970.

11. Mori, T. and Kato, S., Effects of orally administered Tween 80 on the mesenteric nodes of intact and adrenalectomized mice, *Tohoku J. Exp. Med.*, 69, 197-209, 1959.

12. Reyniers, J. P., Machado, E. A., and Farkas, W. R., Hepatotoxicity of chemically defined diets containing polysorbate 80 in germfree and conventional mice, in *Germfree Research. Proceedings of the VIII International Symposium on Germfree Research*, Wostmann, B. S., Ed., Alan R. Liss, New York, 1985, 91.

13. Pleasants, J. R., Johnson, M. H., and Wostmann, B. S., Adequacy of chemically defined, water-soluble diet for germfree BALB/c mice through successive generations and litters, *J. Nutr.*, 116, 1949, 1966.

14. Levenson, S. M., Kan, D., Gruber, C., Jaffe, E., Nakao, K., and Seifter, E., Role of cysteine ethyl ester and indigenous microflora in the pathogenesis of an experimental hemolytic anemia, azotemia, and pancreatic acinar atrophy, in *Germfree Research: Biological Effect of Gnotobiotic Environments. Proceeding of the IV International Symposium on Germfree Research*, Heneghan, J. B., Ed., Academic Press, New York, 1973, 297.

15. Reyniers, J. A. and Sacksteder, M. R., Observations on the survival of germfree C3H mice and their resistance to a contaminated environment, *Proc. Anim. Care Panel*, 8, 41, 1958.

16. Pleasants, J. R., Reddy, B. S., Zimmerman, D. R., Bruckner-Kardoss, E., and Wostmann, B. S., Growth, reproduction and morphology of naturally born, normally suckled germfree guinea pigs, *Z. Versuchstierk.*, 9, 195, 1967.

17. Reddy, B. S., Pleasants, J. R., Zimmerman, D. R., and Wostmann, B. S., Iron and copper utilization in rabbits as affected by diet and germfree status, *J. Nutr.*, 87, 189, 1965.

18. Hashimoto, K., Handa, H., Umehara, K., and Sasaki, S., Germfree mice reared on an "antigen-free" diet, *Lab. Anim. Sci.*, 28, 38, 1978.

19. Pleasants, J. R., Bruckner-Kardoss, E., Bartizal, K. F., Beaver, M. H., and Wostmann, B. S., Reproductive and physiological parameters of germfree C3H mice fed chemically defined diet, in *Recent Advances in Germfree Research. Proceedings of the VIIth International Symposium on Gnotobiology*, Sasaki, S. et al., Eds., Tokai University Press, Tokyo, 1981, 333.

20. Kellogg, T. F. and Wostmann, B. S., Stock diet for colony production of germfree rats and mice, *Lab. Anim. Care*, 19, 812, 1969.

21. Schultze, M. O., Nutrition of rats with compounds of known chemical structure, *J. Nutr.*, 61, 585, 1957.

22. Pleasants, J. R. and Wostmann, B. S., Superior reproduction by BALB/c vs. C3H/He-Cr germfree mice fed chemically defined, ultrafiltered, antigen-free diet, in *Germfree Research: Microflora Control and its Application to the Biomedical Sciences. Proceedings of the VIII International Symposium on Germfree Research*, Wostmann, B. S., Ed., Alan R. Liss, New York, 1985, 87.

23. Wostmann, B. S. and Pleasants, J. R., The germfree animal fed a chemically defined diet: A unique tool, *Proc. Soc. Exp. Biol. Med.*, 198, 539, 1991.

24. Rose, W. C., Coon, M. J., and Lambert, G. F., The amino acid requirements of man. VI. The role of caloric intake, *J. Biol. Chem.*, 210, 331, 1954.

25. Itoh, H., Kishi, T., and Chibata, I., Comparable effects of casein and amino acid mixtures simulating casein on growth and food intake in rats, *J. Nutr.*, 103, 1709, 1973.

26. Yamanaka, M., Nomura, T., Tokioka, J., and Kametaka, M., A comparison of the gastrointestinal tract in germfree and conventional mice fed an amino acid mixture or purified whole-egg protein, *J. Nutr. Sci. Vitaminol.*, 26, 435, 1980.

27. Morin, C. L., Ling, V., and Bourassa, D., Small intestinal and colonic changes induced by a chemically defined diet, *Dig. Dis. Sci.*, 25, 123, 1980.

28. Reddy, B. S., Pleasants, J. R., and Wostmann, B. S., Effect of intestinal microflora on calcium, phosphorus and magnesium metabolism in rats, *J. Nutr.*, 99, 353, 1969.

29. Wostmann, B. S., Pleasants, J. R., and Bealmear, P., Dietary stimulation of immune mechanisms, *Fed. Proc.*, 30, 1779, 1971.

30. Wostmann, B. S., Pleasants, J. R., Bealmear, P., and Kincade, P. W., Serum proteins and lymphoid tissues in germfree mice fed a chemically defined, water soluble, low molecular diet, *Immunology*, 19, 443, 1970.

31. Pleasants, J. R., Schmitz, H. E., and Wostmann, B. S., Immune responses of germfree C3H mice fed chemically defined, ultrafiltered "antigen-free" diet, in *Recent Advances in Germfree Research. Proceedings of the VIIth International Symposium on Gnotobiology*, Sasaki, S. *et al.*, Eds., Tokai University Press, Tokyo, 1981, 535.

32. Hooijkaas, H., Benner, R., Pleasants, J. R., and Wostmann, B. S., Isotypes and specificities of immunoglobulins produced by germfree mice fed chemically defined ultrafiltered "antigen-free" diet, *Eur. J. Immunol.*, 14 1127, 1984.

33. Hooijkaas, H., van der Linde-Preesman, A. A., Bitter, W. M., Benner, R., Pleasants, J. R., and Wostmann, B. S., Isotype expression and specificity repertoire of lipopolysaccharide-reactive B cells in germfree mice fed chemically defined, ultrafiltered "antigen-free" diet, in *Germfree Research. Proceedings of the VIIIth International Symposium on Germfree Research*, Wostmann, B. S., Ed., Alan R. Liss, New York, 1985, 359.

34. Hooijkaas, H., van der Linde-Preesman, A. A., Bitter, W. M., Benner, R., Pleasants, J. R., and Wostmann, B. S., Frequency analysis of functional immunoglobulin C- and V-gene expression by mitogen-reactive B cells in germfree mice fed chemically defined ultrafiltered "antigen-free" diet, *J. Immunol.*, 134, 2223, 1985.

35. Bos, N. A., Benner, R., Wostmann, B. S., and Pleasants, J. R., "Background" Ig-secreting cells in pregnant germfree mice fed a chemically defined ultrafiltered diet, *J. Reprod. Immunol.*, 9, 237, 1986.

Chapter VII

IMMUNOLOGY, INCLUDING RADIOBIOLOGY AND TRANSPLANTATION

GENERAL ASPECTS

The first comprehensive review of the immunological aspects of GF life can be found in a 1959 volume of the *Annals of the New York Academy of Science*, for years referred to by workers in the field as THE BIBLE.[1] Being a part of the proceedings of a meeting held in New York in 1958, it covered most of what was known about GF animals at the time.

The various papers presented at the meeting extended the morphological observations of Glimstedt, who first described the smaller lymphoid nodes of the GF guinea pig,[2] to those of the GF rat, mouse, and chicken[3] and added extensive histological observations.[4,5] They also gave the first quantitative electrophoretic description of the various serum globulin fractions of the GF rat, mouse, and chicken[5-7] in addition to describing the low but variable levels of bacterial agglutinins in the GF animal.[8] Furthermore, they described the generally somewhat slower blood clearance of suspended carbon and of radiolabeled heat-killed bacteria in GF mice[9] (see also Reference 10). The latter observation complemented an earlier statement that phagocytic activity of circulating leukocytes from GF rats was lower for certain, but not for all types of bacteria (Doll, J. P., personal communication).

Taken together, the data implied a more limited and less active lymphoid tissue in the GF animal, as suggested by histological observation and a lower serum gamma globulin fraction. Olson and Wostmann[11] found that the mesenteric lymph nodes of CV Swiss-Webster mice contained on an absolute basis more than ten times more blast cells and potential antibody-producing cells than their GF counterparts. Miyakawa *et al.*[12] studied Peyer's patches (PPs) in GF and CV Wistar rats. They found a similar number along the small intestine, but state that the patches were smaller and "underdeveloped". Pabst *et al.*,[13] who studied postnatal development of PPs in pigs, reported essentially similar differences between GF and CV animals, although in this species PPs occurred in discrete patches only in the jejunum and upper ileum but as a long continuous patch in the terminal ileum. A later study by Bélisle et al.[14] revealed that, in comparison with the lymph nodes of CV Sprague-Dawley rats, the cortex of the cervical and mesenteric nodes of the GF rats were underdeveloped, whereas brachial, inguinal, and popliteal nodes were comparable. However, from the beginning the question remained: "To what extent was the immune potential of the GF animal lastingly impaired?"

Excellent reviews of the earlier studies have been published by Bealmear[15] and Bealmear et al.[16]

EARLY STUDIES: IMMUNE POTENTIAL OF GERMFREE RATS AND MICE MAINTAINED ON SOLID DIETS

Antibody-Forming Potential

A first answer to the previous question came from a study by Gustafsson and Laurell.[17] They concluded that upon association of the GF rat with a CV microflora the gamma globulin level eventually reached a value normal for the CV animal, but only after a pronounced lag phase. A similar observation was reported by Wostmann and Gordon.[18] They found that after the physical introduction of a cecal content suspension of CV rats, it took about 4 weeks for the electrophoretically determined gamma globulin levels to reach values found in CV rats.

A much more severe challenge was given to the GF rat when it was orally inoculated with an 18-h culture of *Salmonella typhimurium ND 750A* presented with the diet. Wostmann[19] reported that this caused the bacterial count to reach approximately $10^9/g$ cecal content within 24 h. All rats lost weight initially, but surprisingly few died. After 3 to 4 weeks the surviving but now GN rats were back to normal weight and had regained all outward appearances of normal, healthy animals.

The spleen and the ileocecal node started to increase in weight within 24 h after inoculation. Both reached a maximum at 10 to 11 days, and declined thereafter to almost preassociation values. During day 2, 3, and 4 it was possible to culture viable salmonella from liver and spleen. After some delay, both gamma globulin levels and agglutinin concentrations rose fast during the first 12 days (Figure 1), the gamma globulin levels thereafter rising more slowly to eventually reach "conventional" values. The immunoelectrophoresis pattern showed an arc in the fast gamma range, normally almost invisible in the patterns of both GF and CV rats, which now became very prominent at day 4 only to regress to its usual appearance by day 10. Carbon clearance values, however, increased 50- to 100-fold already by day 6, to return to their original value 24 days after association (Figure 1). Obviously, the GF rat, with its undersized reservoir of immunocompetent cells, relies on its phagocytic ability as a first line of defense that can be mobilized before its antibody-producing potential can take over. The arc in the fast gamma range which had become prominent on the second day but had regressed at day 10, apparently represents a stress protein possibly related to the early, very pronounced increase in phagocytic potential demonstrated by the carbon clearance values. Recently, Emoto et al.[20] reported that upon association of GF mice with *S. cholerae-suis* heat shock protein p65 was abundantly expressed by macrophages, while Baumann and Gauldie[21] have documented acute phase proteins originating in the liver.

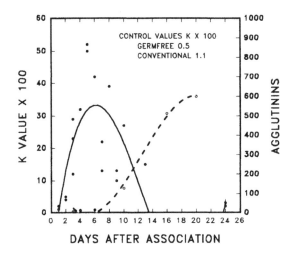

FIGURE 1
Carbon clearance (●) and hemagglutinin values (○) in ex-germfree Lobund-Wistar rats associated with *Salmonella typhimurium*. Phagocytic index K as described in Reference 19.

Although it had become obvious that after association with a CV microflora, with its variety of stimulatory products, GF rats and mice would eventually attain a normal range of immune globulins, the matter of the more quantitative aspects of the immune potential of the GF animal was still considered unanswered. In hindsight the problem rather was whether GF animals, deprived from birth of the continuous stimulation by a viable microflora and its products, would still have the potential to respond adequately to antigenic stimulation given its poorly developed lymphoid system. This question would be especially important in the case of the GF mice maintained for generations on a chemically defined, low molecular, virtually antigen-free diet (CD diet; see Chapter VI) where only the barest minimum of nonmicrobial exogenic stimulus could be expected.

Stimulation with the more defined antigens 7S HGG or *S. typhimurium* vaccine showed that cells from the cervical and mesenteric nodes of GF mice reacted in a manner similar to their CV counterparts (Figure 2).[22] In both GF and CV mice plasmacytopoiesis and lymph node hyperplasia proved to be more pronounced when they were stimulated with the bacterial antigen. However, in response to each antigen, plasmacytic cells, blast cells, and large lymphocytes often showed a proportionately even greater increase in GF than in CV mice, suggesting that even if the GF mice originally had fewer competent cells these cells might be less committed due to lack of previous stimulation. Similar data had been reported by Bauer et al.[23] and Horowitz et al.[24] Baker and Landy[25] stimulated GF and CV BALB/c mice with purified bacterial antigens and at first observed a somewhat delayed response in the spleen of the GF animals. Thereafter, however, antigen-reactive cells increased at the same rate as in their CV counterpart, similar to the aforementioned data represented in Figure 2. Bosma et al.[26] stimulated GF mice with sheep red blood cells, an antigen chosen specifically because their CV mouse strain had shown a very low response and the corresponding GF mice virtually no natural hemagglutinin response. They then evaluated the immune competence of dispersed spleen cells in terms of hemagglutinin and hemolysin formation, and found GF and CV mice comparable in numbers of immunocompetent progenitor cells and in the kinetics of hemagglutinin and hemolysin formation at various ages. A number of years later Seibert et al.[27] investigated the immune potential of GF and CV SJL/J mice because this mouse develops a Hodgkin's disease-like syndrome of presumed viral origin at a time when its immune potential shows a dramatic decline. They found essentially no difference in anti-SRBC plaque-forming response or in IgM or IgG_1 formation, which all started to decline after 4 months of age. The characteristic pathology became obvious after 8 months and proved to be more severe in CV than in GF mice (see also Chapter IX).

Similar results were obtained with rats by McDonald et al.[28] who actually found that the responses to skin allografts, sheep red blood cells, and in the mixed lymphocyte reaction were actually somewhat more pronounced in the GF animal. This similarity between GF and CV rats was again confirmed by Nielsen.[29] Thus, even by the 1960s it had become clear that the lymphocyte population, and notably the plasma cell line, of the GF animal was low in quantity, but quite adequate in quality.

Phagocytosis

The generally observed initial delay in antibody formation may be fatal, however, in certain strains of GF mice that do not have the potential for fast mobilization of phagocytic ability demonstrated by the GF Lobund-Wistar rat (L-W rat).[19] These GF rats withstood oral inoculation with *S. typhimurium* well because of an almost instantaneous strong increase in the phagocytic ability of the liver and spleen before actual antibody formation could afford much further protection. Nardi et al., on the

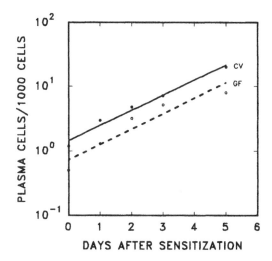

FIGURE 2

Plasma cell formation in mesenteric lymph nodes of germfree and conventional Swiss-Webster mice stimulated with 7S HGG. From data by Olson and Wostmann.[22]

other hand, have reported that oral inoculation with *S. enteritidis* was uniformly fatal for GF CFW mice, whereas all CV animals survived.[30] They attribute an obvious delay in protective IgG formation for the outcome, which clearly suggests that this mouse strain could not develop the phagocytic potential needed for survival during the early phases of the challenge. It has been suggested that phagocytic ability is an early aspect of the defense mechanism in general, and as such the rat seems better equiped than the mouse. However, this and other differences in the quantitative aspects of the immune response between the various studies cited above might also be explained by differences in experimental protocol. Especially, the diet may have been a factor since the antigenic load caused by the dead bacteria in the sterilized diets must have varied substantially with the composition of the diet. Outzen *et al.* showed that resistance to infection in mice increased with age, presumably because of the more extended exposure to the antigenicity of diet and bedding.[31,32]

The phagocytic response is affected mainly by three factors: the phagocytic ability of the cell per se, the availability of opsonizing specific or nonspecific proteins, and the availability of complement. To our knowledge no difference between GF and CV animals has been reported in the availability of complement factors. Opsonizing proteins such as agglutinins may be found in the serum of the GF animal, depending largely on the materials used in the diet and on the sterilization protocol. Wagner has shown that casein is a notorious source of dead microbial forms (Table 1; from Wostmann *et al.*[33]). Later, Hammer and Hingerle[34] reported substantial influence of food antigens on what they termed "preformed natural antibodies" in GN dogs and pigs, but at the time it was not quite clear to what extent the GF state would affect the phagocytic potential of these cells per se.

TABLE 1
Bacterial Count on Casein Samples By
Direct Microscopy[a,b]

Casein	Number of individual bacterial cells per gram
Hammersten	6.70×10^8
Vitamin-free casein[c]	1.31×10^9
Soluble casein, Na caseinate	5.34×109

[a] Data from Wostmann et al.[33]
[b] Modified Breed method.
[c] General Biochemicals, Inc.

Thorbecke and Benaceraf had found clearance of an i.v. injected carbon suspension, standardized for the weight of liver plus spleen, slightly lower in GF mice.[9] Later, Lefevre et al.[35] reported on the uptake of per os-administered latex particles by the PPs in GF and CV mice. Whereas the number and distribution of PPs did not differ significantly, the CV mice showed significantly larger follicles and accumulated correspondingly more latex particles, suggesting comparable uptake ability of the individual phagocytic cells. Even when "opsonized" sheep red blood cells were used as a target, no difference in the engulfing potential of peritoneal macrophages could be detected between GF and CV mice of the same strain.[36]

There appears to be agreement on the adequacy of the potential of the phagocyte of the GF animal to engulf foreign materials, a capacity which would obviously be enhanced by the presence of an opsonizing antibody.[37] However, Bauer et al.[38] reported that phagocytosis is not only dependent on the presence of an opsonizing antibody produced as a response to microbial antigens, but also on the digestive capability of the phagocyte which may be enhanced by the microflora. They concluded that a slower digestion of antigenic material would lead to a delayed presentation of immunogenic fragments, thereby causing an initial delay in antibody formation. However, this same slower digestion could also lead to a more sustained response.[39]

Actually, the whole phagocytic process appears to depend on recruitment of the phagocytosing cell, its attachment to the offending matter, the subsequent internalization of the material, and its eventual destruction by digestion. Most workers find that macrophages of GF rats and GF mice are slow to respond to chemotactic stimuli. Johnson and Balish[40] have reported that GF BALB/c mice were less able, after an oral load of *Listeria monocytogenes*, to prevent accumulation of viable organism in the spleen. They ascribed this to a slower mobilization of blood-borne monocytes, since the functional ability of the macrophages was found to be comparable to that of the CV mice. A similar observation had been reported earlier by Jungi and McGregor in GF rats.[41] Morland et al.,[42] going into further detail and comparing Fc-receptor- and C3b-receptor-mediated phagocytosis in vitro concluded that the peritoneal macrophages of GF and CV mice were comparable in many ways. However, unstimulated β-glucuronidase activity of the lysates of GF macrophages was increased, whereas spreading on glass, the aforementioned chemotatic response, in vitro induction of lysosomal enzymes, and the capacity to internalize via the C3b receptor after stimulation were reduced or absent. Conventionalization of the mice by contact with CV mice brought all values to CV levels within 4 weeks.[43] A study in rats by Johnson and Balish[44] led to similar conclusions about the peritoneal macrophages, but found the tumoricidal activity of the pulmonary alveolar macrophages of GF mice to be much less responsive to lipopolysaccharide (LPS) stimulation than those of their CV counterparts.

Late in 1981, in a follow-up study of the pulmonary alveolar macrophage, Starling and Balish[45] summarized (and referenced) the then available data on the immune function of the GF animal as follows:

> The absence of a microbial flora in GF animals is associated with an increased susceptibility to infectious agents and an ability to generate only a feeble delayed-type hypersensitivity reaction compared to that of the CV animal. GF animals do, however, manifest T lymphocyte function and are capable of synthesizing Igs to thymus-dependent and -independent Ags, and rejecting skin allografts. Compared to CV animals, macrophages of GF animals have been shown to be defective in several functions: phagocytosis via C3b receptors, chemotactic responsiveness, and lysosomal enzyme activity.

Their further study of the alveolar macrophage led to the general conclusion that the function of these macrophages was enhanced, both in terms of proliferation and in terms of lysosomal activity, by the presence of an intestinal microflora. The lower direct tumor cell cytotoxicity of the GF rat[45] would then be explained by a lower lysosomal enzyme activity. A few years later Czuprynski and Brown,[46] in a study involving nude mice of the BALB/c strain, looked at the poor performance of GF mice after inoculation with L. monocytogenes. Their in vitro studies demonstrated that the functional deficit was with the macrophages rather than with the neutrophils. The data indicated, as before, that intracellular killing rather than defective phagocytosis was to blame, although no defect in the production of the superoxide anion (O_2^-) could be established. Mitsuyama et al.,[47] on the other hand, report the O_2^- production by phorbol myristate-stimulated peritoneal macrophages from GF ICR mice to be always significantly lower than that of their comparable CV counterparts. They consider that aspect of macrophage function as definitely being influenced by the microflora.

Recently, Talafantora and Mandel[48] reported that GF piglets associated with *Streptococcus faecalis* were protected against the lethal effects of *Escherichia coli* O 55 which otherwise killed the monoassociated piglets within 24 h. When they studied the concentration of active phagocytes in the circulation of the GF piglets they found hardly an increase during the first year of life. Monoassociation with *S. faecalis* dramatically increased the number of circulating granulocytes and active phagocytes within 24 h. Association with *E. coli* O 83 or a mixture of anaerobic bacteria, on the other hand, did not cause any major change in phagocytic ability.[49] Kamijo *et al.*[50] have pointed to the fact that, next to γ-interferon, LPS is the most potent activator of the NO synthase, providing nitric oxide as a cytotoxic agent to the macrophage.

All data thus far seem to point at the GF macrophage as defective in chemotaxis and in destructive and lytic capacity, with its concomitant potential of the production and presentation of antigenic material. Lately, a lower concentration of plasma fibronectin, a glycoprotein with opsonic properties, has been added to the deficiencies in the GF mouse potentially causing lower phagocytic activity.[51] However, as the earlier-mentioned association study of GF L-W rats with *S. typhimurium* had demonstrated, even the phagocytically disadvantaged GF rat could mount a quite adequate life-saving phagocytic response.[19]

B CELLS, T CELLS, AND NK CELLS

As mentioned before, quite early on it had become obvious that the lymph nodes of the GF animal, especially those in direct contact with the gastrointestinal tract, were small and underdeveloped[2,3] with few active germinal centers.[5] In the absence of a viable intestinal microflora and its products (e.g., LPS), dietary components, dead bacteria, and possibly animal-to-animal contact between non-inbred animals are the only potential sources of exogenous antigenic stimulation for the GF animal. However, once exposed to a specific antigenic stimulation, the antibody-forming potential of cells isolated from these nodes or from the spleen proved to be as good, and often even slightly better, than of those isolated from comparable CV organs.[22,24-26,28,29] Translated into more modern terms, the GF animal starts with less of a functional B cell population in the critical lymphoid organs; this notwithstanding the fact that data obtained by Bos *et al.* seem to indicate that exogenous antigenic stimulation makes little difference in the background level of B cell production in the bone marrow.[52] However, those strategically positioned B cells proved to be eminently competent — upon antigenic challenge of the GF animal, in most instances only a modest delay in antibody production would occur. This response, although generally quite adequate, might be somewhat lower and slower than in the CV animal,[22] presumably due to the absence of adjuvant stimulatory materials otherwise produced by the intestinal microflora. Studies with GF mice suggest that for certain strains at least, in case of a severe microbial challenge this delay in effective antibody production may be fatal.[30]

It had been realized since the early 1960s that the thymus plays a key role in the ontogeny of the immune response. Soon it was recognized as the maturation site of that most important branch of the immune system: the T cell population with its subclasses of helper, repressor, and cytotoxic cells. While the B cell population remains quite stable with age, T cell populations were found to decrease,[53] presumably in relation to the slow involution of the thymus. Along with the thymus, the immune potential of the older animal appears to decline. No differences in thymus weight[54,55] or cellularity[56] have been found between GF and CV rats at various ages. In the GF mouse thymus growth patterns are again very similar, and gross appearances are the same. However, Wilson et al. reported that the thymus in the GF mouse grows somewhat slower and is generally somewhat smaller than in the CV mouse.[57] A similar observation has been reported by van der Waay.[58]

Data obtained in the Lobund Aging Study indicate that in the GF rat WBC values are, on the average, about 75% of those found in their CV counterparts. Studying the Lobund material, Gilman-Sachs et al. found little difference in the percentage of circulating T lymphocytes populations between GF and CV rats,[59] although the percentage of cytotoxic/suppressor cells was higher, in GF animals of all age groups. Earlier, Vos, comparing T cell populations of the thymus, spleen, and lymph nodes of GF BALB/c mice raised on the aforementioned CD diet (see Chapter VI), again had found no significant differences with the CV BALB/c mice. From his studies he concluded that the available T cells were fully functional, would produce cytokinins, and could support isotype switching.[60]

In recent years it has been recognized that the T cell population consists of two apparently rather independent lines, the $\alpha\beta$-cells and the $\tau\delta$-cells, depending on the structure of the T cell receptor. It would appear that $\tau\delta$-T cells function as a first line of defense against infectious pathogens, since relatively large numbers are found at epithelial surfaces that face a potentially hostile environment.[61] Bandiera et al. have reported that in the intestinal epithelium of the mouse the presence or absence of a normal microbial flora appeared not to affect the $\gamma\delta$-T cell population.[62] Umesaki et al.[63] recently showed that in the intestinal epithelium of GF BALB/c mice $\gamma\delta$-T cells significantly outnumbered $\alpha\beta$-T cells. Upon conventionalization as described, the $\gamma\delta$-T cell population hardly changed, but the $\alpha\beta$-T cells increased sharply to a level several times greater than the relatively stable $\gamma\delta$-T cell numbers. This augments the observation by Knight and Wostmann[64] that the lymphoid cells in the lamina propria of the GF rat numbered only 20% of those in the CV animal. This can now be presumed to largely be a deficit in $\alpha\beta$-T cells. Apparently the $\gamma\delta$-T cells are a fixture which functions as a first line of defense, but αβ-T cell numbers are enhanced by the presence of an intestinal microflora, possibly via an extrathymic pathway.[65]

Bealmear et al.[16] have summarized the numerous data on mitogen stimulation of the spleen cells of GF and CV mice. Although different workers sometimes come to somewhat different conclusions, depending on strain, diet, and methodology used, the overall impression is that T cell mitogen stimulation (PHA and ConA) of cells from GF and CV mice give largely comparable results. Table 2 shows data taken from a recent study by Kiyono et al. using BALB/c mice.[66] They again show a possibly even higher response in the GF animals. The above conclusion appears to hold true even for GF mice maintained on the virtually antigen-free CD diet. A study using C3H mice showed that spleen lymphocytes from GF animals fed either commercial diet L-485 (see Chapter V) or CD diet and stimulated with PHA or ConA again resulted in almost identical proliferation, although in this case the absolute values were slightly lower than those obtained with CV mice lymphocytes.[67] The Kiyono et al. data show, on the other hand, that B cell stimulation by LPS appears to result in almost twice the proliferation rate in GF as in CV mice[16] (Table 2). All in all, the data appear to indicate that the immune systems of GF rats and GF mice are, in any case qualitatively, and after some initial delay also quantitatively, at least as effective as that of their CV counterparts. The presence of a microbial flora in the gut serves to maintain a state of readiness, depending on the nature of the microflora and its metabolic products. Here we think not only of the well-known action of LPS,[68] but also include the stimulating action of materials like bacterially derived peptidoglycan-polysaccharide polymers, as described recently by Woolverton et al.[69]

TABLE 2
Splenic Lymphocyte Proliferative Response of GF
and CV BALB/c Mice Maintained on Commercial Diet[a]

Stimulant	Concentration (mg/ml)	GF response[b]	CV response
None		1,002 ± 182	927 ± 157
PHA	1	55,129 ± 4,480	42,231 ± 3,620
ConA	0.5	69,297 ± 4,230	54.078 ± 4,735
PWM	10	29,365 ± 2,797	22,000 ± 2,963
Ph-LPS	10	28,534 ± 1,573	16,435 ± 1,311

[a] Data with permission from Kiyono et al.[66]
[b] Data expressed as cpm 3H-TdR incorporated/8 × 105 cells.

A third cell population considered of major importance in the immunological surveillance of the body is the natural killer or NK cell. These are cells in the progenitor line of the T cells, but which have not been processed by passage through the thymus.[70] Their aim appears to be the cytotoxic removal of virally infected or otherwise malignant cells. In apparent accordance with this function, their numbers increase with age. However, their quantitative evaluation appears to be affected by the specifics of the methodology used for their determination. One is left with the impression that, in general, their numbers and/or cytotoxic effectiveness in GF animals compares well with those of their CV counterparts. Tazume et al.[71] reported higher NK activity in the spleen of GF than of CV Sprague-Dawley rats, whereas Gilman-Sachs et al. found slightly, but not significantly higher numbers of circulating NK cells in GF L-W rats.[59] Data by Bartizal et al.[72] suggest comparable NK activity of the spleen cells of GF and CV C3H/HeCr mice, but show that when these GF mice had been raised on an antigen-free diet (CD diet, see Chapter VI), NK activity of the spleen cells increased by an average of around 100% in young, and by about 40% in older mice (Table 3). At the same time the data indicate that age dependence had disappeared in the CD diet group. In this strain, absence of exogenous antigenic stimulation apparently allowed early full development of NK potential. On the other hand, working with BALB/c mice, and obviously using the same methodology, the authors report much lower NK activity in the GF animals.[72] Then again, Susuki et al. appear to find similar activity values in their strain of GF and CV BALB/c mice.[73] However, from these and other data one does get the impression that the presence of certain microflora elements may repress NK activity.

TABLE 3

The Effect of Age, Diet, and Microbial Status on the
Tumoricidal Activity of Spleen Cells From C3H/HeCr Mice[a]

Status	Diet[b]	Age (wk)	Mean% Cytotoxicity at E:T Ratio		
			25:1	50:1	100:1
Germfree	CD	6–10	10.0	12.1	16.6
Germfree	L-485		4.3	5.2	8.2
Conventional	L-485		3.6	5.2	9.3
Germfree	CD	29–36	8.8	13.2	16.3
Germfree	L-485		6.8	8.9	11.1
Conventional	L-485		4.7	6.4	9.5

[a] Data with permission from Bartizal et al.[72]

[b] CD: chemically defined, see Chapter VI; L-485: natural ingredient diet, see Chapter V.

IMMUNE GLOBULINS

During the 1950s Grabar and Courcon[74] had continued to develop immunoelectrophoresis, a technique which could visualize over 20 protein fractions in serum. This and related techniques made it possible to distinguish the various classes of immune globulins, indicated as IgM, IgA, and the various IgGs. Analyzing the serum of GF and CV L-W rats, Grabar et al.[75] found IgM, only small amounts of a rather fast-moving IgG, and no trace of IgA in the GF animals. Crabbé et al. and Benveniste et al. confirmed the absence of IgA in the serum of GF mice[77-79] and showed that it takes at least several weeks of exposure to a "normal" microflora before IgA starts to appear in the serum. However, stimulation by an orally administered antigen-like ferritin would readily trigger IgA formation in the PPs of GF mice, the IgA being of the secretory type, consisting of the IgA dimer combined with the secretory component. The secreted form is presumed to protect the intestinal mucosa and, once triggered in the gut by antigenic stimulation, this potential for IgA production then translates to other sites to protect exposed epithelium. Eventually the internal nodes will also start producing and secreting IgA into the blood. This IgA is in the monomeric form, and its function in the circulation is not well understood. However, obviously an antigenic challenge of the intestinal PPs, preferably of microbial origin,[80] is needed to start the process. A paper by Kawanishi et al. seems to imply that a bacterial product like LPS may be a factor in the transition of B cells from IgM to IgA production.[81] Later, Vos,[60] studying adult GF BALB/c mice that had been raised on the virtually antigen-free chemically defined diet (CD diet, Chapter VI), could confirm that under those circumstances some IgG, but no IgA secreting cells could be detected in the spleen. Although Vos states that GF BALB/c mice raised under such conditions contain fully functional T cells capable of secreting the lymphokines needed for isotype switching, other stimulatory factors appear to be needed to bring this about. Recently IL-4 and IL-7, besides LPS, have been have mentioned in this regard.[82]

An intriguing observation has been reported by Durkin et al.[83] Like others, they found smaller and presumably fewer PPs in GF than in CV rats; but these workers found that in the PPs of 8- to 10-week-old GF Sprague-Dawley rats maintained on autoclaved Purina rat chow, the B cell population contained up to 38% IgA surface-bearing (sIgA) cells against only 19% in comparable CV rats and that about half of these also carried sIgE. Conventionalization of these animals, or exposure to bacterial cell wall components like peptodiglycans, but not to LPS, within 18 h brought the B cell population of the PPs to a composition seen in the CV control animals, with only a trace of sIgE-bearing cells. Here we have to bear in mind that LPS may not be as much of a mitogenic agent for the rat as it is for the mouse.[84] One may speculate that in this case specific components in the autoclaved diet may have brought about this rather unusual situation in the GF rats, but no definite explanation can be given.

Another interesting observation has been reported by Satoh et al.[85] They found no IgA-containing plasma cells in the ileal lamina propria of GF rats, but did so in the Golgi area of the crypt and villus epithelium, where IgA appeared to be localized in the Paneth cells. Rodning et al.[86]

had speculated earlier that Paneth cells might be involved in the regulation of the intraluminal microflora within the crypt region. Satoh *et al.* are not certain about the origin of IgA in the Paneth cells, but speculate that redirection of already discharged and subsequently reabsorbed IgA may be involved. The diet is not mentioned in their publication, but its composition may explain a discrepancy that seems to exist between their data and those of others.

Benner *et al.*[87] report that whereas mature GF and CV BALB/c mice show comparable numbers of background IgM- and IgG-producing cells in the spleen, IgA-producing cells in the spleen of these GF mice comprised only a few percent of the numbers found in their CV counterparts. Later studies would confirm that GF rodents always show substantial amounts of serum IgM, even in GF mice maintained on a chemically defined, virtually antigen-free diet.[33,88] Thus, whereas serum IgM appeared to be an intrinsic expression of the immune system, the various IgG components, and especially IgA, needed exogenous stimulation to express themselves. For this reason IgM has often been equated with "natural antibody".

NATURAL ANTIBODY

In the meantime, workers at the Pasteur Institute had come to recognize that the immune potential may have an "autonomous core" of functional B and T cells. In mice this "core" would be represented mainly by the spleen, and would maintain itself by recognition of self constituents — possibly in the process of being degraded — and would therefore be largely independent of exogenous stimulation.[89,90] This reservoir of mostly IgM-producing cells could then react to "overlapping" exogenous antigen, and adjust via hypermutation to meet the challenge of these antigens. The spleen appeared to be the main source of this circulating IgM, often considered to be "natural antibody", but studies comparing GF-CD and CV BALB/c mice revealed that at least part of this "natural antibody" in the CV mice was caused by stimulation by carbohydrate antigens of microbial origin.[91]

CYTOKININS

More information has recently become available about these most important factors, the cytokines, which to a large extent govern the interplay of the various elements of the immune system. Granholm *et al.* thus far have not found any difference in the amount or tissue distribution of IL-1 between GF and CV rats.[92] The same group tested the proliferative response of thymocytes isolated from GF and CV NMRI mice to IL-1α, IL-1β, and IL-6 in the tritiated thymidine incorporation test run with suboptimal amounts of phytohemagglutinin, and found that the GF mice showed about half the response of their CV counterparts.[93] A later study appears to confirm the higher cytokine responsiveness of the CV NMRI mouse, suggesting that exposure to microflora induces a higher state of activation of the immune system in that strain.[94] This observation seems to be at variance with the rather general agreement mentioned earlier in this chapter that mitogen responses of the spleen cells of GF mice are at least comparable to and sometimes better than those of their CV counterparts. Extending their studies on the ontogeny of IL-1 in rats, Granholm and co-workers[95] concluded that IL-1 activity appears at a definitive time point in fetal life, and that exposure of rats to microbial antigens or microbial products does not affect its production or bioactivity.

RADIATION BIOLOGY AND BONE MARROW TRANSPLANTATION

Because of the necessity for the use of radiation to be able to establish allogeneic and possibly even xenogeneic chimeras, and in general to make organ transplantation possible, it was important to establish the influence of a viable microflora on such procedures. As early as 1956 Reyniers *et al.* had reported that GF rats survived approximately twice as long as CV rats following exposure to single doses of whole-body irradiation of between 300 and 1000 rad.[96] The early studies have been summarized by van Bekkum,[97] and show that in the dose range of the bone marrow syndrome GF mice tolerate, on the average, around 100 rad more than their CV counterparts. In the range of the gastrointestinal syndrome, death starts at 1500 rad for CV and around 2000 rad for GF mice, while in the range above 2000 rad GF mice tend to live twice as long as the CV mice. Largely similar data have been reported by McLaughlin *et al.*[98] and Walburg *et al.*[99]

Onoue *et al.*[100] exposed GF and monoassociated mice to 2 krad whole-body gamma irradiation and reported average survival time for the GF mice at 10.8 days, against CV mice of around 7 days. Of the bacterial species tested only *E. coli* and a *Pseudomonas* sp. brought the survival time down to the CV range, while a *Clostridium* sp. increased survival time to about 14 days. The radiation procedure proved to have no obvious effect on the number of viable associated bacteria excreted with the feces. It is not quite clear which aspect of the monoassociates determined their positive or negative action. However, Wilson *et al.*[101] have thought in terms of a toxic substance, possibly an endotoxin, that would be released by the microflora during irradiation and would interfere with iron metabolism via a loss of transferrin from the circulation. The above-mentioned studies did indicate that the intestinal microflora had an uncontrolled, mostly negative effect on the pathological sequelae following damaging levels of irradiation. The GF animal therefore presented itself as a model of choice for the study of radiation damage and organ transplantation.

In the years after World War II the possibility of nuclear attack created the necessity to understand as much as possible about radiation pathology. It soon became obvious that bone marrow transplantation would play an important role in repairing the milder forms of radiation damage. At the same time this procedure might eventually prove to be a cure for various forms of leukemia, especially if ways could be found to use bone marrow from not quite matching donors.

The possibility of reconstituting GF mice with allogeneic bone marrow (CFW to C3H, and C3H to CFW) was first explored by Connell and Wilson,[102] who studied a radiation range between 700 and 1600 rad. Reconstitution was checked by ascertaining that the donor RBCs had been established. They reported substantially increased survival of the reconstituted animals, especially in the GF mice, and stressed the fact that "the microflora of an animal contributes to the syndrome known as secondary disease." In a later study by Wilson and co-workers, GF C3H mice were irradiated with 1000 rad and reconstituted with DBA/2 bone marrow. Since CV mice had shown a lower radiation tolerance, the CV C3H mice received only 825 rad but were reconstituted with the same dose of DBA/2 bone marrow. Whereas the GF mice showed 98% survival at 120 days, the CV mice showed only 64% at 30 days, and no survival at 120 days. The latter animals showed all the signs of GvH disease described in the literature, but in the GF mice this was described as "rare", with a total absence of the usually observed diarrhea.[103]

In 1985 Pollard and associates reported the first successful establishment of long-lived mouse-to-rat chimeras in the GF state, and especially emphasized that "no evidence of GvH disease was seen over a long period."[104,105] More extensive data were reported in 1987, when 18 irradiated GF L-W rats, reconstituted with bone marrow from GF CFW mice and who had survived for more than 3 months, were available for study. Again true chimerism could be established. Low to normal hematocrit and WBCs were found, and circulating immunoglobulins were normal for the GF state. NK cell activity was depressed, as were the mitogen stimulation results. The data suggested that after this xenogeneic transplantation, reconstitution, both hematologically and immunologically, was incomplete. However, it was specifically stated that no acute GvH reaction was seen.[106] The same group later extended this research by including gnotobiotic rats populated by an intestinal microflora consisting of *Streptococcus faecalis*, *Lactobacillus acidophilus*, *L. brevis*, *Bifidobacterium animalis*, and *Clostridium bifermentans*. All GF and GN L-W rats received 900 rad, and were reconstituted with either syngeneic or allogeneic (Sprague-Dawley) rat bone marrow, or with xenogeneic GF CFW mouse bone marrow. In this series all syngeneic and allogeneic chimeras survived until they were sacrificed 4 months later. Even of the xenogeneic mouse-to-polyassociated rat chimeras, three out of four survived for that length of time, a survival rate comparable to that of the GF mouse/GF rat chimeras mentioned earlier, but under the polyassociated condition these rats did show signs of GvH disease. All showed normal NK cell activity, except again the GN chimeras which showed no activity. T cell mitogenic response was decreased in the allogeneic, and more so in the xenogeneic chimeras, especially in the presence of the above pentaflora. In general, however, the Gram-positive pentaflora had limited effect on the performance indicators of the rat-mouse chimera.[107] Veenendaal et al. would later claim that the anaerobic intestinal microflora plays a major role in the development of GvH disease.[108] Heidt and Vossen,[109] in a recent study, point to elements of the anaerobic microflora of the recipient presumably not present in the donor mice as activating donor T cells to induce GvH disease.

It would seem, however, that the matter of presence or absence of GvH reaction in the GF state, especially in the case of xenogeneic chimeras, is not resolved to everybody's satisfaction. Much may depend on the strain or substrain of donor and recipient. For a more complete review of the earlier studies see Bealmear et al.[110]

IMMUNOLOGICAL EVALUATION OF GF MICE MAINTAINED ON A CHEMICALLY DEFINED, LOW MOLECULAR WEIGHT ANTIGEN-FREE DIET

The development of the chemically defined "antigen-free" diet described in Chapter VI, together with the elimination of possible animal-to-animal antigenicity through the use of inbred mice, produced IgA and IgG levels undetectable by immunoelectrophoresis even in 14-month-old GF-CD C3H mice. It decreased the already low WBC counts of the GF mice even further. Only IgM values remained in the range of those of solid diet-fed GF and CV mice.[33] Hashimoto et al.[111] fed a similar CD diet formulation to GF ICR mice and used antisera specific for the different Ig isotypes to quantitate serum levels in 5- to 6-week-old animals. Again IgM levels were only slightly decreased in the absence of antigenic exposure. IgA was not detected in either the serum or intestinal wall of any GF mice, whatever their diet; it was detected only in mice with a viable microflora. The serum IgG level of GF-CD mice was 1/10 that of GF-NI mice, and 1/100 that of CV-NI mice. The authors concluded that IgG levels were sensitive to both dietary and microbial antigens. Later, Bos et al.[91] would verify these data in greater detail, finding at least a trace of serum IgA in CD diet-fed BALB/c mice.

However, after immunization using sheep red blood cells (SRBC), GF-CD C3H mice at the Lobund Laboratory showed an immune potential as least as good as that of GF-NI or CV-NI mice. This was expressed in their postinoculation levels of anti-SRBC plaque-forming cells, serum IgG,[33] and hemagglutinin.[112] Following *in vitro* challenge with phytohemagglutinin and concanavalin A, the spleen cell response of GF-CD mice was again at least equal to that of GF and CV mice fed the NI diet.[112] In the nonchallenged state, spleens of 8- to 10-week-old GF-CD mice showed IgM-containing cells in numbers equal to those in comparable CV-NI spleens, but only 20% as many IgG-containing cells, and no IgA-containing cells.[60]

Study of the "spontaneously" occurring ("background") Ig synthesis in spleen and bone marrow of C3H mice by Hooijkaas *et al.*[88] again revealed roughly equal numbers of IgM-secreting cells in older GF-CD and CV-NI mice, although the numbers in 2-month-old GF-CD mice tended to be lower. The number of IgG-secreting cells in the spleen was comparable in both age groups, but was found to be lower in the bone marrow of the GF-CD mice. IgA-secreting cells were drastically decreased. The antibody-specificity repertoire of the background IgM-secreting cells when tested against five differently haptenized SRBCs was found comparable between the two animal groups, suggesting that the generation of the IgM antibody repertoire was independent of external antigenic stimulation.

In a follow-up study the isotype expression and the specificity repertoire of LPS-reactive B cells was studied. LPS-activated cultures of spleen and bone marrow cells of GF-CD C3H mice produced similar, or possibly slightly higher, numbers of IgM- and IgG_1-producing clones than the cells of their CV counterparts fed NI diet L-485. Again, the specificity repertoire of the stimulated cells appeared not to be influenced by prior antigenic exposure of the intact animals.[113] The study also suggested that the capability of IgM-secreting cells to switch to secretion of IgG_1 was not affected by prior antigen exposure.[114]

More extensive study by Bos *et al.*[115] of the "background" antigen-specific spleen and bone marrow cells of 2-month-old GF-CD BALB/c mice largely confirmed results obtained earlier with the C3H strain. In GF-CD mice, *in vitro* IgM-secreting cells outnumbered IgG- and especially IgA-producing cells in both organs. Similar results have also been reported by Vos *et al.*[60] In the comparable BALB/c CV mice, IgM-secreting cells were dominant in the spleen but not in the bone marrow. As was found in the case of the C3H mice, these IgM-producing cells seemed to occur in a somewhat higher concentration in the GF-CD than in the CV-NI mice. Comparison of the frequency distribution of Ig-secreting cells specific for DNP27-BSA and the anti-idiotype monoclonal antibodies Ac38 and Ac146 in spleen and bone marrow of GF-CD and CV mice suggested a shift in the distribution of the IgG- and IgA-producing cells specific for these antigens when compared with the quite similar antigen-specific distribution of the IgM-secreting cells (Table 4). This and other experiments indicate that while the IgM antibody repertoire is relatively independent of exogenous antigenic stimulation, the shift from IgM to IgG and IgA production appears to be influenced by antigenic exposure. This suggests two compartments of background Ig-secreting cells: a stable, endogenously regulated compartment consisting mainly of IgM-secreting cells, and another compartment consisting mainly of IgG- and IgA-secreting cells whose numbers and specificity repertoire are affected by exogenous antigenic stimulation.

TABLE 4
Frequency of DNP27-BSA-Specific Ig-Secreting Cells in the Spleen and Bone Marrow of Germfree BALB/c Mice Fed Chemically Defined (CD) Diet, and of Conventional BALB/c Mice Fed Autoclaved Natural Ingredient (L-485) Diet[a]

		Frequency[b]	
	Ig-Secreting Cells	**GF-CD**	**CV-L-485**
Spleen	IgM	1 in 143	1 in 137
	IgG	1 in 48	1 in 1766
	IgA	1 in 93	1 in 3433
Bone marrow	IgM	1 in 30	1 in 21
	IgG	1 in 71	1 in 1112
	IgA	1 in 113	1 in 228

[a] Data taken from Table 3 in Bos et al.[115]
[b] Frequency in terms of DNP27-BSA-specific Ig-secreting cells.

Serum Ig levels of the GF-CD mice correlate well with the numbers of Ig-secreting cells of the various isotypes in the spleen of these mice, while in CV mice there exists a clear discrepancy. In the serum of CV-NI mice IgG is the dominant isotype, while in the spleen IgM-secreting cells are clearly in the majority. This difference cannot be explained by the difference in half-life of the isotypes, and suggests that serum Ig in GF-CD mice reflects the production of the spleen, while in CV-NI mice organs such as bone marrow and lymph nodes contribute considerably to the serum Ig levels, notably for IgG and IgA.[91]

The development of the compartment of Ig-secreting cells was studied in GF-CD and CV-NI mice from birth to young adult age. The results suggest that the ontogenetic appearance of Ig-secreting cells in the spleen and the specificity repertoire of the IgM-secreting cells are largely independent of exogenous antigenic stimulation. However, after birth the rate of development of the Ig-secreting cell compartments was enhanced by environmental antigenic stimulation.[116]

CONCLUSION

The GF rat and especially the GF mouse have proven to be excellent models for the study of the many factors that determine immune potential.

In the absence of exogenous stimulation the full potential remains in the background, especially as far as B cell function and phagocytosis are concerned. Exogenous stimulation is needed to enlarge B cell populations, to promote isotype switching, to enhance phagocytic capability, and possibly to enhance cytokine response. As mentioned earlier, Vos et al.[60] found the T cells of the GF-CD mouse fully functional, including the production of cytokines and their support for isotype switching.

REFERENCES

1. Reyniers, J. A., Ed., Germfree vertebrates: present status, Ann. N.Y. Acad. Sci., 78, 1959.
2. Glimstedt, G., Bakterienfreie Meerschweinchen. Afzucht, Lebensfähigkeit und Wachstum, nebst Untersuchungen über das lymphatische Gewebe, Acta Physiol. Microbiol. Scand., Suppl. 30, 1, 1936.
3. Gordon, H. A., Morphological and histological characterization of germfree life, in Germfree Vertebrates: Present Status, Reyniers, J. A., Ed., Ann. N.Y. Acad. Sci., 78, 208, 1959.
4. Miyakawa, M., The lymphatic system of germfree guinea pigs, Ann. N.Y. Acad. Sci., 78, 221, 1959.
5. Thorbecke, J., Some histological and functional aspects of lymphoid tissue in germfree animals. I. Morphological studies, Ann. N.Y. Acad. Sci., 78, 237, 1959.
6. Wostmann, B. S., Serum proteins in germfree vertebrates, Ann. N.Y. Acad. Sci., 78, 255, 1959.
7. Gustafsson, B. E. and Laurell, C., Gamma globulin in germfree rats, J. Exp. Med., 108, 251, 1958.
8. Wagner, M., Serological aspects of germfree life, in Germfree Vertebrates: Present Status, Reyniers, J. A., Ed., Ann. N.Y. Acad. Sci., 78, 261, 1959.
9. Thorbecke, G. J. and Benaceraf, B., Some histological and functional aspects of lymphoid tissue in germfree animals. II. Studies on phagocytosis in vivo, Ann. N.Y. Acad. Sci., 78, 247, 1959.
10. Doll, J. P., Rate of carbon clearance in three strains of germfree mice, Am. J. Physiol., 203, 291, 1962.
11. Olson, G. B. and Wostmann, B. S., Lymphocytopoiesis and cellular proliferation in nonantigenically stimulated germfree mice, J. Immunol., 97, 267, 1966.
12. Miyakawa, M., Sumi, Y., Sakurai, K., Ukai, M., Hirabayashi, N., and Ito, G., Serum gamma-globulin and lymphoid tissue in the germfree rats, Acta Haematol. Jpn., 32, 501, 1969.
13. Pabst, R., Geist, M., Rothkötter, H. J., and Fritz, F. J., Postnatal development and lymphocyte production of jejunal and ileal Peyer's patches in normal and gnotobiotic pigs, Immunology, 64, 539, 1988.
14. Bélisle, C., Sainte-Marie, G., and Peng, F., Tridimensional study of the deep cortex of the rat lymph node. VI. The deep cortex units of the germfree rat, Am. J. Pathol., 107, 70, 1982.
15. Bealmear, P.M., Host defense mechanisms in gnotobiotic animals, in Immunologic Defects in Laboratory Animals, Vol. 2, Gershwin, E. and Marchant, B., Eds., Plenum Press, New York, 1981, 261.
16. Bealmear, P. M., Holtermann, O. A., and Mirand, E. A., Influence of the microflora on the immune response, in The Germfree Animal in Biomedical Research, Coates, M. E. and Gustafsson, B. E., Eds., Laboratory Animal Handbooks 9, Laboratory Animals Ltd., London, 1984, 335.

17. Gustafsson, B. E. and Laurell, C., Gamma globulin production in germfree rats after bacterial contamination, *J. Exp. Med.*, 110, 675, 1959.

18. Wostmann, B. S. and Gordon, H. A., Electrophoretic studies on the serum protein pattern of the germfree rat and its changes upon exposure to a conventional bacterial flora, *J. Immunol.*, 84, 27, 1960.

19. Wostmann, B. S., Antimicrobial defense mechanisms in the *Salmonella typhimurium* associated ex-germfree rat, *Proc. Soc. Exp. Biol. Med.*, 134, 294, 1970.

20. Emoto, M., Naito, T., Nakamura, R., and Yoshikai, Y., Different appearance of gamma delta T cells during salmonellosis between Ityr and Itys mice, *J. Immunol.*, 150, 3411, 1993.

21. Baumann, H. and Gauldie, J., The acute phase response, *Immunol. Today*, 15, 74, 1994.

22. Olson, G. B. and Wostmann, B. S., Cellular and humoral immune response of germfree mice stimulated with 7S HGG or *Salmonella typhimurium*, *J. Immunol.*, 97, 275, 1966.

23. Bauer, H., Horowitz, R. E., Levenson, S. M., and Popper, H., The response of the lymphatic tissue to the microbial flora. Studies on germfree mice, *Am. J. Pathol.*, 42, 471, 1963.

24. Horowitz, R. E., Bauer, H., Paronetto, F., Abrams, G. D., Watkins, K. C., and Popper, H., The response of lymphatic tissue to bacterial antigen. Studies in germfree mice, *Am. J. Pathol.*, 44, 747, 1964.

25. Baker, P. J. and Landy, M., Brevity of the induction phase of the immune response of mice to capsular polysaccharide antigens, *J. Immunol.*, 99, 687, 1967.

26. Bosma, M. J., Makinodan, T., and Walburg, H. E., Development of immunologic competence in germfree and conventional mice, *J. Immunol.*, 99, 420, 1967.

27. Seibert, K., Pollard, M., and Nordin, A., Some aspects of humoral immunity in germfree and conventional SJL/J mice in relation to age and pathology, *Cancer Res.*, 34, 1707, 1974.

28. McDonald, J. C., Zimmerman, G., and Bollinger, R. R., Immune competence of germfree rats. I. Increased responsiveness to transplantation and other antigens, *Proc. Soc. Exp. Biol. Med.*, 136, 987, 1971.

29. Nielsen, H. E., Reactivity of lymphocytes from germfree rats in mixed leukocyte and graft-*versus*-host reaction, *J. Exp. Med.*, 136, 417, 1972.

30. Nardi, R. M., Vieira, E. C., Crocco-Alfonso, L. C., Silva, M. E., Andrade, A. M. V., Bambirra, E. A., and Nicoli, J. R., Experimental salmonellosis in conventional and germfree mice: bacteriological and immunological aspects, *Microecol. Ther.*, 20, 313, 1990.

31. Outzen, H. C. and Pilgrim, H. I., Differential mortality of male and female C3H mice introduced in a conventional colony, *Proc. Soc. Exp. Biol. Med.*, 124, 52, 1967.

32. Outzen, H. C., Ageing and resistance to infection in germfree C3H mice, in *Germfree Biology*, Mirand, E. A. and Back, N., Eds., Plenum Press, New York, 1969, 207.

33. Wostmann, B. S., Pleasants, J. R., and Bealmear, P., Dietary stimulation of immune mechanisms, *Fed. Proc.*, 30, 1779, 1971.

34. Hammer, C. and Hingerle, M., Development of preformed natural antibodies in gnotobiotic dogs and pigs; impact of food antigens on antibody specificity, *Transplant. Proc.*, 24, 707, 1992.

35. Lefevre, M. E., Joel, D. D., and Schidlovsky, G., Retention of ingested latex particles in Peyer's patches of germfree and conventional mice, *Proc. Soc. Exp. Biol. Med.*, 179, 522, 1985.

36. Perkins, E. H., Nettesheim, P., Morita, T., and Walburg, H. E., The engulfing potential of peritoneal phagocytes of conventional and germfree mice, in *The Reticuloendothelial System and Atherosclerosis*, Deluzio, N. R. and Paoletti, R., Eds., Plenum Press, New York, 1967, 175.

37. Trippestad, A. and Midtvedt, T., The phagocytic activity of polymorphonuclear leucocytes from germfree and conventional rats, *Acta Pathol. Microbiol. Scand.*, 79, 519, 1971.

38. Bauer, H., Horowitz, R. E., Watkins. K. C., and Popper, H., Immunologic competence and phagocytosis in germfree animals with and without stress, *J. Am. Med. Assoc.*, 187, 715, 1964.

39. Bauer, H., Paronetto, F., Burns, W. A., and Einheber, A., The enhancing effect of the microbial flora on macrophage function and the immune response. A study in germfree mice, *J. Exp. Med.*, 123, 1013, 1966.

40. Johnson, W. J. and Balish, E., Macrophage function in germfree, athymic (nu/nu), and conventional-flora (nu/+) mice, *J. Reticuloendothel. Soc.*, 28, 55, 1979.

41. Jungi, T. W. and McGregor, D. D., Impaired chemotactic reponsiveness of macrophages from gnotobiotic rats, *Infect. Immun.*, 19, 553, 1978.

42. Morland, B., Smievoll, A. I., and Midtvedt, T., Comparison of peritoneal macrophages from germfree and conventional mice, *Infect. Immun.*, 26, 1129, 1979.

43. Morland. B. and Midtvedt, T., Phagocytosis, peritoneal influx, and enzyme activities in peritoneal macrophages from germfree, conventional and ex-germfree mice, *Infect. Immun.*, 44, 750, 1984.

44. Johnson, W. J. and Balish, E., Direct tumor cell and antibody-dependent cell-mediated cytotoxicity by macrophages from germfree and conventional rats, *J. Reticuloendothel. Soc.*, 29, 205, 1981.

45. Starling, J. R. and Balish, E., Lysosomal enzyme activity in pulmonary alveolar macrophages from conventional, germfree, monoassociated and conventionalized rats, *J. Reticuloendothel. Soc.*, 30, 497, 1981.

46. Czuprynski, C. J. and Brown, J. F., Phagocytes from flora-defined and germfree athymic nude mice do not demonstrate enhanced antibacterial activity, *Infect. Immun.*, 50, 425, 1985.

47. Mitsuyama, M., Ohara, R., Amako, K., Nomoto, K., Yokokura, T., and Nomoto, K., Ontogeny of macrophage function to release superoxide anion in conventional and germfree mice, *Infect. Immun.*, 52, 236, 1986.

48. Talafantová, M. and Mandel, L., Protective activity of *Streptococcus faecalis* against pathogenic action of *Escherichia coli* O 55 in gnotobiotic piglets, *Folia Microbiol.*, 30, 329, 1985.

49. Mandel, L., Talafantová, M., Trebichavsky, I., and Trávnícek, J., Stimulating effect of *Streptococcus faecalis* on phagocytosis in gnotobiotic piglets, *Folia Microbiol.*, 34, 68, 1989.

50. Kamijo, R., Harada, H., Matsuyama, T., Bosland, M., Gerecitano, J., Shapiro, D., Le, J., Koh, S. I., Kimura, T., Green, S. J., Mak, T. W., Tanaguchi, T., and Vilcek, J., Requirement for transcription factor IRF-1 in NO synthase induction in macrophages, *Science*, 263, 1612, 1994.

51. Anderlik, P., Szeri, I., Banos, Z. S., Barna, Z. S., and Kalabay, L., Effects of microorganisms and their substances on the plasma fibronectin concentration in gnotobiotic mice, *Microecol. Ther.*, 20 395, 1990.

52. Bos, N. A., Pleasants, J. R., Wostmann, B. S., Freitas, A. A., and Benner, R., Renewal rate of B cells in germfree BALB/c mice fed a chemically defined diet, *Microecol. Ther.*, 20, 191, 1990.

53. Yoshikai, Y., Matsuzaki, G., Kishihara, K., Nomoto, K., Yokokura, T., and Nomoto, K., Age-associated increase in the expression of T cell antigen receptor gamma-chain gene in conventional and germfree mice, *Infect. Immun.*, 56, 2069, 1988.

54. Gordon, H. A., Bruckner-Kardoss, E., Staley, T. E., Wagner, M., and Wostmann, B. S., Characteristics of the germfree rat, *Acta Anat.*, 64, 367, 1966.

55. Klausen, B. and Hougen, H. P., Quantitative studies of lymphoid organs, blood and lymph in inbred athymic and euthymic LEW rats under germfree and specific-pathogen-free conditions, *Lab. Anim.*, 21, 342, 1987.

56. Uno, Y., Sumi, Y., and Sakura, K., Studies on the thymus of germfree rats, in *Advances in Germfree Research and Gnotobiology. Proceedings of the International Symposium on Germfree Life Research, Nagoya, Japan*, Miyakawa, M. and Luckey, T. D., Eds., CRC Press, Boca Raton, FL, 1967, 129.

57. Wilson, R., Bealmear, M., and Sobonya, R., Growth and regression of the germfree (axenic) thymus, *Proc. Soc. Exp. Biol. Med.*, 118, 97, 1965.

58. Van der Waay, D., Influence of the intestinal flora on the relative thymus weight, in *Gnotobiology and Its Applications. Proceedings of the IX International Symposium on Gnotobiology*, Ed. Fond. Marcel Marieux, Lyon, France, 1987, 281.

59. Gilman-Sachs, A., Kim, Y. B., Pollard, M., and Snyder, D. L., The influence of aging, environmental antigens, and dietary restriction on expression of lymphocyte subsets in germfree and conventional Lobund-Wistar rats, *J. Gerontol.*, 46, B101, 1991.

60. Vos, Q., Jones, L. A., and Kruisbeek, A. M., Mice deprived of exogenous antigenic stimulation develop a normal repertoire of functional T cells, *J. Immunol.*, 149, 1204, 1992.

61. Born, W. K., O'Brien, R. L., and Modlin, R. L., Antigen specificity of γδ T lymphocytes, *FASEB J.*, 5, 2699, 1991.

62. Bandiera, A., Monta-Santos, T., Itohata, S., Degermann, S., Heusser, C., Tonegawa, S., and Coutinho, A., Localization of γδ T cells is independent of normal microbial colonization, *J. Exp. Med.*, 172, 239, 1990.

63. Umesaki, Y., Setomaya, H., Matsumoto, S., and Okada, Y., Expansion of αβ T cell receptor-bearing intestinal intraepithelial lymphocytes after microbial colonization in germfree mice, and its independence from thymus, *Immunology*, 79, 32, 1993.

64. Knight, P. L. and Wostmann, B. S., Influence of *Salmonella typhimurium* on ileum and spleen morphology of germfree rats, *Indiana Acad. Sci.*, 74, 78, 1964.

65. Takeuchi, M., Miyazaki, H., Mirokawa, K., Yokokura, T., and Yoshikai, Y., Age-related changes of T cell subsets in the intestinal intraepithelial lymphocytes of mice, *Eur. J. Immunol.*, 23, 1409, 1993.

66. Kiyono, H., McGhee, J. R., and Michalek, S. M., Lipopolysaccharide regulation of the immune response: comparison of responses to LPS in germfree *Escherichia coli*-monoassociated and conventional mice, *J. Immunol.*, 124, 36, 1980.

67. Pleasants, J. R., Schmitz, H. E., and Wostmann, B. S., Immune responses of germfree C3H mice fed chemically defined ultrafiltered "antigen-free" diet, in *Recent Advances in Germfree Research. Proceedings of the VII International Symposium on Gnotobiology*, Sasaki, S. *et al.*, Eds., Tokai University Press, Tokyo, 1981, 535.

68. Moreau, M. C., Hudault, S., and Bridonneau, C., Systemic antibody response to ovalbumin in gnotobiotic C3H/HeJ mice associated with *Bifidobacterium bifidum* or *Escherichia coli*, *Microecol. Ther.*, 20, 309, 1990.

69. Woolverton, C. J., Holt, L. C., and Sartor, R. B., Oral peptidoglycan-polysaccharide stimulates systemic immunocompetency in germfree mice, *Microb. Ecol. Health Dis.*, 7, 183, 1994.

70. Lanier, L. L., Spits, H., and Phillips, J. H., The developmental relationship between NK cells and T cells, *Immunol. Today*, 13, 392, 1992.

71. Tazume, S., Wade, A., and Pollard, M., Natural killer cell activity in germfree and conventional rats, in *Germfree Research: Microflora Control and its Application in the Biomedical Sciences. Proceedings of the VIII International Symposium on Germfree Research*, Wostmann, B. S., Ed., Alan R. Liss, New York; *Prog. Clin. Biol. Res.*, 181, 343, 1985.

72. Bartizal, K. F., Salkowski, C., Pleasants, J. R., and Balish, E., The effect of microbial flora, diet, and age on the tumoricidal activity of natural killer cells, *J. Leukocyte Biol.*, 36, 739, 1984.

73. Susuki, T., Susuki, Y., Maruo, N., Yamada, T., Katigiri,S., Kawamura, E., Okano, T., Kubota, Y., Ghoda, A., Yamashita, R., Shimizu, M., and Nomoto, K., Immunoresponses and natural killer cell activity of nude and normal mice under conventional and germfree conditions, in *Germfree Research: Microflora Control and its Application to the Biomedical Sciences. Proceedings of the VIII International Symposium on Germfree Research*, Wostmann, B. S., Ed., Alan R. Liss, New York; *Prog. Clin. Biol. Res.*, 181, 325, 1985.

74. Grabar, P. and Courcon, J., Étude des serums de cheval, lapin, rat et souris par l'analyse immunoélectrophorétique, *Bull. Soc. Chim. Biol.*, 40, 1993, 1958.

75. Grabar, P., Courcon, J., and Wostmann, B. S., Immunoelectrophoretic analysis of the serum of germfree rats, *J. Immunol.*, 88, 679, 1962.
76. Wostmann, B. S., Recent studies on the serum proteins of germfree animals, *Ann. N.Y. Acad. Sci.*, 94, 272, 1961.
77. Crabbé, P. A., Bazin, H., Eyssen, H., and Heremans, J. F., The normal microbial flora as a major stimulus for the proliferation of plasma cells synthesizing IgA in the gut, *Int. Arch. Allerg.*, 34, 362, 1968.
78. Benveniste, J., Lespinats, G., Adam, C., and Salomon, J.-C., Immunoglobulins in intact, immunized, and contaminated axenic mice: study of serum IgA, *J. Immunol.*, 107, 1647, 1971.
79. Benveniste, J., Lespinats, G., and Salomon, J.-C., Serum and secretory IgA in axenic and holoxenic mice, *J. Immunol.*, 103, 1656, 1971.
80. Nagura, H., Hasegawa, H., Yoshimura, S., Aihara, K., Watanabe, K., Sawamura, S., and Ozawa, A., Comparative immunohistochemical studies on conventional and germfree rat intestinal mucosa: with special reference to microbial flora and secretory IgA (sIgA), in *Recent Advances in Germfree Research. Proceedings of the VII International Symposium on Gnotobiology*, Sasaki, S. et al., Eds., Tokai University Press, Tokyo, 1981, 511.
81. Kawanishi, H., Saltzman, L., and Strober, W., Mechanisms regulating IgA class-specific immunoglobulin production in murine gut-associated lympoid tissues. II. Terminal differentiation of postswitch sIgA-bearing Peyer's patch B cells, *J. Exp. Med.*, 158, 649, 1983.
82. Lydyard, P. M., Kearney, J. F., Burrows, P., and Gathings, W., Lymphopoiesis today, *Immunol. Today*, 15, 255, 1994.
83. Durkin, H. G., Chise, S. M., Gaetjens, E., Bazin, H., Tarcsay, L., and Dukor, P., The origin and fate of IgE-bearing lymphocytes. II. Modulation of IgE isotype expression on Peyer's patch cells by feeding with certain bacteria and bacterial cell wall components or by thymectomy, *J. Immunol.*, 143, 1777, 1989.
84. Wells, C. L. and Balish, E., The mitogenic activity of lipopolysaccharide for spleen cells from germfree, conventional and gnotobiotic rats, *Can. J. Microbiol.*, 25, 1087, 1979.
85. Satoh, Y., Ishikawa, K., Tanaka, H., and Ono, K., Immunohistochemical observations of immunoglobulin A in the Paneth cells of germfree and formerly-germfree rats, *Histochemistry*, 85, 197, 1986.
86. Rodning, C. B., Erlandsen, S. L., Wilson, I. D., and Carpenter, A., Light microscopic morphometric analysis of rat ileal mucosa. II. Component quantitation of Paneth cells, *Anat. Rec.*, 204, 33, 1982.
87. Benner, R., van Oudenaren, A., Björklund, M., Ivars, F., and Holmberg, D., 'Background' immunoglobulin production: measurement, biological significance and regulation, *Immunol. Today*, 3, 243, 1982.
88. Hooijkaas, H., Benner, R., Pleasants, J. R., and Wostmann, B. S., Isotypes and specificities of immunoglobulins produced by germfree mice mice fed chemically defined ultrafiltered "antigen-free" diet, *Eur. J. Immunol.*, 14, 1127, 1984.
89. Pereira, P., Forni, L., Larsson, E., Cooper, M., Heusser, C., and Coutinho, A., Autonomous activation of B and T cells in antigen-free mice, *Eur. J. Immunol.*, 16, 685, 1986.
90. Avrameas, S., Natural autoantibodies: from "horror autotoxicus" to "gnothi seauton", *Immunol. Today*, 12, 154, 1991.
91. Bos, N. A., Kimura, H., Meeuwsen, C. G., de Visser, H., Hazenberg, M. P., Wostmann, B. S., Pleasants, J. R., Benner, R., and Marcus, D. M., Serum immunoglobulin levels and naturally occurring antibodies against carbohydrate antigens in germfree BALB/c mice fed chemically defined ultrafiltered diet, *Eur. J. Immunol.*, 19, 2335, 1989.
92. Granholm, T., Midtvedt, T., and Söder, O., Lymphocyte activating cytokinins in different tissues from germfree and conventional rats, *Microecol. Ther.*, 20, 329, 1990.

93. Söder, O., Granholm, T., Lundström, C., Fröysa, B., and Midtvedt, T., Lymphocyte responsiveness to cytokinins in germfree, ex-germfree and conventional mice, *Microecol. Ther.*, 20, 197, 1990.

94. Granholm, T., Froysa, B., Lundstrom, C., Wahab, A., and Midtvedt, T., Cytokine responsiveness in germfree and conventional NMRI mice, *Cytokine*, 4, 545, 1992.

95. Granholm, T., Froysa, B., Midtvedt, T., and Soder, O., Ontogeny of lymphocyte activating factors in conventional and germfree rats, *Reg. Immunol.*, 4, 209, 1992.

96. Reyniers, J. A., Trexler, P. C., Scruggs, W., Wagner, M., and Gordon, H. A., Observations on germfree and conventional albino rats after total-body X-irradiation, *Radiat. Res.*, 5, 591, 1956.

97. van Bekkum, D. W., Radiation biology, in *The Germfree Animal in Research*, Coates, M. E. *et al.*, Eds., Academic Press, London, 1968, 257.

98. McLaughlin, M. M., Dasquisto, M. P., Jacobus, J. P., and Horowitz, R. E., Effects of the germfree state on responses of mice to whole-body irradiation, *Radiat. Res.*, 23, 333, 1964.

99. Walburg, H. E., Mynatt, E. I., and Robie, D. M., The effect of strain and diet on the thirty-day mortality of X-irradiated germfree mice, *Radiat. Res.*, 23, 333, 1964.

100. Onoue, M., Uchida, K., Yokokura, T., Takahashi, T., and Mutai, M., Effect of intestinal microflora on the survival time of mice exposed to lethal whole-body gamma irradiation, *Radiat. Res.*, 88, 533, 1981.

101. Wilson, R., Barry, T. A., and Bealmear, P. M., Evidence of a toxic substance of bacterial origin in the blood of irradiated mice, *Radiat. Res.*, 41, 89, 1970.

102. Connell, S. J. and Wilson, R., The treatment of x-irradiated germfree CFW and C3H mice with isologous and homologous bone marrow, *Life Sci.*, 4, 721, 1965.

103. Jones, J. M., Wilson, R., and Bealmear, P. M., Mortality and gross pathology of secondary disease in germfree mouse radiation chimeras, *Radiat. Res.*, 45, 577, 1971.

104. Pollard, M., Protected environment and its utility in experimental allogeneic and xenogeneic bone marrow transplantation, *Tokai J. Exp. Clin. Med.*, 10, 175, 1985.

105. Pollard, M., Luckert, P. H., and Meshorer, A., Xenogeneic bone marrow chimerism in germfree rats, in *Germfree Research: Microflora Control and its Applications to the Biomedical Sciences. Proceedings of the VIII International Symposium on Germfree Research*, Wostmann, B. S., Ed., Alan R. Liss, New York, 1985, 447.

106. Wade, A. C., Luckert, P. H., Tazume, S., Neidbalski, J. L., and Pollard, M., Characterization of xenogeneic mouse to rat chimeras. Examination of hematologic and immunologic function, *Transplantation*, 44, 88, 1987.

107. Eberly, K., Bruckner-Kardoss, E., and Wagner, M., Immunological effects of a Gram positive pentaflora in rat radiation chimeras, in: *Gnotobiology and its Applications. Proceedings of the IX International Symposium on Gnotobiology*, Ed. Fond. Marcel Mérieux, Lyon, France, 1987, 172.

108. Veenendaal, D., de Boer, F., and van der Waaij, D., Effect of selective decontamination of the digestive tract of donor and recipient on the occurrence of murine delayed-type graft vs. host disease, *Med. Microbiol. Immunol.*, 177, 133, 1988.

109. Heidt, P. J. and Vossen, J. M., Experimental and clinical gnotobiotics: influence of the microflora on graft-versus-host disease after allogeneic bone marrow transplantation, *J. Med.*, 23 161, 1992.

110. Bealmear, P. M., Holterman, O. A., and Mirand, E. A., Radiation pathology and treatment, in *The Germfree Animal in Biomedical Research*, Coates, M. E. and Gustafsson, B. E., Eds., Laboratory Animal Handbooks 9, Laboratory Animals Ltd., London, 1984, 413.

111. Hashimoto, K., Handa, H., Umehara, K., and Sasaki, S., Germfree mice reared on an "antigen-free" diet, *Lab. Anim. Sci.*, 28, 38, 1978.

112. Pleasants, J. R., Schmitz, H. E., and Wostmann, B. S., Immune responses of germfree C3H mice fed chemically defined ultrafiltered "antigen-free" diet, in *Recent Advances in Germfree Research. Proceedings of the VII International Symposium on Gnotobiology*, Sasaki, S. *et al.*, Eds., Tokai University Press, Tokyo, 1981, 535.

113. Hooijkaas, H., van der Linde-Preesman, A. A., Bitter, W. M., Benner, R., Pleasants, J. R., and Wostmann, B. S., Isotype expression and specificity repertoire of lipopolysaccharide-reactive B cells in germfree mice fed chemically defined ultrafiltered "antigen-free" diet, in *Microbial Control and its Application in the Biomedical Sciences. Proceedings of the VIII International Symposium on Germfree Research*, Wostmann, B. S., Ed., Alan R. Liss, New York; *Prog. Clin. Biol. Res.*, 181, 359, 1985.

114. Hooijkaas, H., van der Linde-Preesman, A. A., Bitter, W. M., Benner, R., Pleasants, J. R., and Wostmann, B. S., Frequency analysis of functional immunoglobulin C- and V-gene expression by mitogen-reactive B cells in germfree mice fed chemically defined ultrafiltered "antigen-free" diet, *J. Immunol.*, 134, 2223, 1985.

115. Bos, N. A., Meeuwsen, C. G., Wostmann, B. S., Pleasants, J. R., and Benner, R., The influence of exogenous antigenic stimulation on the specificity repertoire of background immunoglobulin-secreting cells of different isotypes, *Cell. Immunol.*, 112, 371, 1988.

116. Bos, N. A., Meeuwsen, C. G., Hooijkaas, H., Benner, R., Wostmann, B. S., and Pleasants, J. R., Early development of IgG-secreting cells in young of germfree BALB/c mice fed a chemically defined ultrafiltered diet, *Cell. Immunol.*, 105, 235, 1987.

Chapter VIII

PARASITOLOGY

GENERAL ASPECTS

The parasite-associated gnotobiote may be seen as an ideal model to study aspects of parasitology because of the absence of an otherwise interactive, highly uncontrollable factor — the "normal" intestinal microflora. When considering the resulting gnotobiote, three main factors will affect the status of the associated parasite: (1) the functional status of the immune system of the host; (2) the potential of the parasite to adapt to the specific host immune responses, both humoral and cellular; and (3) the environment, nutritional and otherwise, which this gnotobiotic status of the host creates for its commensal. Since many of the functional and metabolic characteristics of the GF rat, GF mouse, and to a lesser extent the GF guinea pig, have been extensively documented, as parasite-associated gnotobiotes they offer a well-controlled model for study. Presently available data indicate that the microflora almost always has a profound influence on the host-parasite relationship. Its absence may prevent the intestinal parasite from becoming established, presumably because of the vastly different conditions in the GF gut (see Chapter III). On the other hand, the less well-developed immune capacity of the GF animal may enhance the parasite's invasive potential, although seldom to the point that the host is overwhelmed. By the same token, parasitic infection of GF animals may be an excellent tool to study many aspects of the development of the immune response.

EFFECTS OF THE MICROFLORA VIA NUTRIENT AVAILABILITY AND THE IMMUNE SYSTEM

Protozoa

One of the earliest studies using GF guinea pigs was carried out by Phillips et al.[1] at the Lobund Laboratory and later at the National Institutes of Health. Using the then available Caesarian-derived GF guinea pigs, they administered an inoculum of Entamoeba histolytica trophozoites

127

intracecally. Whereas 90% of the CV controls developed acute ulcerative amoebic lesions, often with lethal consequences, the GF guinea pigs retained the organisms for only a few days, without any signs of pathology. Guinea pigs monoassociated with either *Escherichia coli* or *Aerobacter aerogenes* developed a pathology similar to that seen in the CV animals, again often with a lethal outcome. In a following study the authors put special emphasis on the altered conditions found in the cecum of the GF guinea pig, and pointed to the much more positive oxidation-reduction potential (–90 vs. –367 mV) in the GF cecum.[2] Later they used an inoculum of *Entamoeba histolytica* cultivated and harvested by procedures designed to enhance its pathogenicity. This resulted in mild lesions in the GF guinea pigs, but in fatal amebiasis in the CV control animals.[3]

Obviously the presence of bacteria proved to be a sine qua non for the development of true ulcerative lesions by *E. histolytica*, presumably because of the special conditions in the GF gut created by the absence of the microflora. On the other hand, when Newton *et al.*[4] introduced an inoculum of *Trichomonas vaginalis* subcutaneously into GF guinea pigs, large lesions and gas formation occurred, whereas the same inoculum produced "little or no evidence of infection" in CV guinea pigs. It would appear that only in a rare case does there seem to be no effect from the presence or absence of a microflora. Thus, one of the critical points in this research is how to obtain the parasite under study free of bacterial or viral contamination without damaging its natural viability. This criterion may invalidate some of the quantitative aspects of the earlier studies. Apparently the early cultures of *E. histolytica* may have been contaminated with hemoflagellates.

Later studies involving parasitic protozoa showed a mixed picture regarding the involvement of the microflora. Owen,[5] working with *Eimeria falciformis*, could not establish the coccid in GF mice after administration of the oocysts by gavage. In the absence of a microflora no penetration of the intestinal epithelium by sporozoites took place. Again, different conditions in the GF gut were blamed. The group led by Vieira, studying parasites of special importance in the Latin American health picture, also found GF mice to be less susceptible to infection with *Leishmania mexicana amazonensis*[6] than their CV controls. They mentioned that the disease is much more severe in CV than in GF mice, but in a recent report Vieira and Scott[7] infected three strains of GF and CV mice with *L. major* and found that the GF 'Swiss' mice used in their studies showed significantly larger lesions 8 weeks after infection than their CV counterparts. They speculate that differences in T helper cell response (Th1 and Th2) and their cytokine products IFN-γ and IL-4 could play a controlling role here. When the bloodforms of the trypomastigotes of *Trypanosoma cruzi* were administered, the developing disease was again more severe in GF than in CV rats and mice.[6] In the latter case the authors thought in terms of the underdeveloped immune system of the GF animal, which apparently cannot provide protection before the parasite is well established; but since more recent data from that group showed that the eventual antitrypanosoma IgG levels in the GF rat were higher than in their comparable CV counterparts,[8] it would once again appear that differences in cellular immunity may play an important role here. Similar thinking resulted from the observation by Harp et al.[9] who could establish Cryptosporidium parvum in GF, but not in CV mice. They suggest the possibility of an insufficient functioning of the T cell system in these GF mice, because nude mice seem unable to clear the parasite once it is established. Recently, Boher et al.[10] pointed especially to the role of the anaerobic bacteria in the ilium, which may contribute to a protecting type of cellular immune response. However, Vos et al.[11] consider the function of the T cells in GF mice in general to be quite adequate (see Chapter VII). Obviously, the role of the immune system in these infections is as yet not well understood, and may vary with the species and strain of the animal model.

Nematodes

Extensive studies of the factors controlling the establishment of the parasite Trichinella spiralis have been carried out by the Polish investigator Przyjalkowski.[12] In an earlier study he had unsuccessfully tried to establish the pinworm Aspiculuris tetraptera in GF mice, an infection easily produced in CV mice. With T. spiralis the crucial role played by the intestinal microflora again became evident. After per os administration of the larvae to GF and CV mice, worm recovery in the intestine of CV mice was 24 times higher and from the muscle 12 times higher than in the GF animal. Mice monoassociated with E. coli showed about a fivefold increase in worm recovery.[13] In a continuation of this approach a number of monoassociates were studied, all with a positive effect on establishment and development of the parasite. This time, worm recovery from the gut of CV mice fed the sterilized Teklad diet* as before, was 13 times higher than from GF mice. However, when an unautoclaved diet was fed to the CV mice the recovery was 57 times as high, implying destruction of nutritional factor(s) by sterilization and the possibility of (nutritional) factors made available by the CV microflora. Later studies also concentrated on immunological factors. Whereas GF mice expelled the worms much easier, their serum IgA and IgG_1 titers were only about half those seen in their CV counterparts.[14,15] The fact that the inflammatory reaction in the CV mice was so much more pronounced than in GF mice may indicate that the parasite, not finding the right (nutritional?) environment in the GF mice, passed through the animal without doing much harm. In the CV mice, on the other hand, the larvae developed and established themselves, giving rise to much increased and possibly damaging

---

* Ralston-Purina Co., St. Louis, MO.

immune reactions.[16] In the case of *T. spiralis* one has the impression that the nutritional factor dominates although the immune system does play a role, especially in the CV animal. Despommier[17] has pointed out that certain secreted antigens of *T. spiralis* will induce a level of protection, while Cabaj[18] suggested the potential involvement of T cells, especially in protecting the CV mouse against lethal consequences of the infection.

Helminths

During the 1960s Wescott[19] had tried to establish the parasitic helminth *Nematospiroides dubius* in GF mice. Again, in the absence of a microflora few of the larvae developed into mature worms, and pathology was minor in contrast to the full-blown infection which developed in the CV mice. A later study indicated somewhat less of a difference, but still showed that in the CV mouse parasite survival and fecal egg count were always higher than in the GF animal. Here again, the authors mention the possibility that the intestinal microflora might provide the parasite with specific nutrients that would favor its development.[20] Weinstein *et al.*,[21] on the other hand, reported almost comparable results in GF and CV mice, with the infection in GF mice overlapping that of the CV mice. In their studies the nematode infection persisted for four to five months in both groups, but the adult worms in the GF host were smaller than normal. This was attributed to the fact that the mice were maintained on a semisynthetic casein-starch formula, instead of on sterilizable lab chow. They found, however, that eggs shed in the GF feces would not develop into normal larvae in that environment unless viable fecal organisms from CV mice had been added. Again, it would seem that nutritional factors may have been involved. The authors point to the possibility that the free-living larvae in feces may actually feed on the live bacteria, in addition to profiting from their metabolic products.

Schistosomiasis and Chagas Disease

Vieira's group of Brazilian workers have studied the mammalian part of the life cycle of *Schistosoma mansoni* in GF and CV mice after obtaining the GF cercariae from GF snails.[22,23] In this case the absence of a viable microflora produced more adult worms and a higher egg recovery in the GF mice than in their CV counterparts. However, higher levels of lethal antibody (as determined according to Tavares *et al.*[24]) were found in the serum of the GF mice. While this suggested a more severe disease in the GF mice, histopathological data indicated more damage in the CV mice. Again, in the case of schistosomiasis the microflora with its multiple effects on the immune potential appeared to play a role that is not quite well understood.

Chagas disease has been, and is still a major health risk in Brazil and other parts of Latin America. Recently, Resende *et al.*[25] from the aforementioned Brazilian group reported the establishment of the complete life cycle of the intracellular parasite *Trypanosoma cruzi* under GF conditions. The Reduviidae insects *(Dipetalogaster maximus)* had been reared under GF conditions, then infected by feeding on GF infected mice. The GF trypomastigotes from the macerated insects were then administered to GF and CV mice. As observed earlier,[6] the pathology caused by this causative agent of Chagas disease was found to be more severe in GF than in CV mice, presumably because the underdeveloped immune system of the GF animal is unable to prevent this parasite from establishing itself. Because of this sensitivity of the GF mouse, this model was then used to investigate the effect of systemic iron levels on the course of the disease.[26] After i.p. injection of 5 mg ferric hydroxide-dextran complex the infected GF and CV mice both showed a substantial increase in parasitemia over untreated infected controls. However, no more difference was seen between the GF and the CV mice, suggesting that the availability of iron may not be a major factor in the development of this disease.

Cestodes

As far as tapeworm infection is concerned, we have been able to find only two studies, both of the genus *Hymenolepis*. Houser and Burns[27] administered the sterile cysticercoids of *H. diminuta* (obtained from infected GN beetle larvae) with the diet to GF and CV rats. Presence or absence of a microflora and its potentially nutritional products had no effect; in both cases the worms developed through a normal life cycle. Earlier, Newton *et al.*[28] had come to the same conclusion studying the fate of *H. nana* in GF and CV guinea pigs.

REFERENCES

1. Phillips, B. P., Wolfe, P. A., Rees, C. W., Gordon, H. A., Wright, W. H., and Reyniers, J. A., Studies on the ameba-bacteria relationship in amebiasis. Comparative results of the intracecal inoculation of germfree, monocontaminated and conventional guinea pigs with *Entamoeba histolytica*, *Am. J. Trop. Med. Hyg.*, 4, 675, 1955.
2. Phillips, B. P., Wolfe, P. A., and Bartgis, I. L., Studies on ameba-bacteria relationships in amebiasis. II. Some concepts on the etiology of the disease, *Am. J. Trop. Med. Hyg.*, 7, 392, 1958.
3. Phillips, B. P., Studies on the ameba-bacteria relationship in amebiasis. III. Induced amebic lesions in the germfree guinea pig, *Am. J. Trop. Med. Hyg.*, 13, 301, 1964.
4. Newton, W. L., Reardon, L. V., and deLeva, A. M., A comparative study of the subcutaneous inoculation of germfree and conventional guinea pigs with two strains of *Trichomonas vaginalis*, *Am. J. Trop. Med. Hyg.*, 9, 56, 1960.

5. Owen, D., *Eimeria falciformis* (Eimer, 1870) in specific pathogen free and gnotobiotic mice, *Parasitology*, 71, 293, 1975.

6. Nicoli, J. R., Vieira, E. C., Moraes-Santos, T., Mota-Santos, T., Bezerra, M., Silva, M. E., Vieira, L. Q., Guimaraes, I. T., Evangelista, F. A., Coello, P. M. Z., Mayrink, W., da Costa, C. A., and Bambirra, F. A., The germfree animal as a model for the study of parasitic diseases: Schistosomiasis mansoni, cutaneous Leishmaniasis and Chagas disease, in *Gnotobiology and its Applications. Proceedings of the IX International Symposium on Gnotobiology*, Ed. Fond. Marcel Mérieux, Lyon, France, 1987, 226 and 265.

7. Vieira, L. Q. and Scott, P., Germfree mice exhibit higher levels of IL-4 than conventional mice when infected with *Leishmania major*, in Proc. XIth Int. Symp. Gnotobiology, Belo Horizonte, Brasil. (Abstr.).

8. Silva M. E., Silva, M. E., Silva M. E., Bambirra, E. A., Nicoli, J. R., and Vieira, E. C., Some parasitological and immunological aspects of the experimental infection with *Trypanosoma cruzi* in germfree and conventional rats, in Proc. XIth Int. Symp. Gnotobiology, Belo Horizonte, Brasil. In press.

9. Harp, J. A., Wannemuehler, M. W., Woodmansee, D. B., and Moon, H. W., Susceptibility of germfree or antibiotic-treated adult mice to *Cryptosporidium parvum*, *Infect. Immun.*, 56, 2006, 1988.

10. Boher, Y., Perez-Schael, I., Caceres-Ditmar, G., Urbina, G., Gonzalez, R., Kraal, G., and Tapia, F. J., Enumeration of selected leukocytes in the small intestine of BALB/c mice infected with *Cryptosporidium parvum*, *Am. J. Trop. Med. Hyg.*, 50, 145, 1994.

11. Vos, O., Jones, L. A., and Kruisbeek, A. M., Mice deprived of exogenous antigenic stimulation develop a normal repertoire of functional T cells, *J. Immunol.*, 149, 1204, 1992.

12. Przyjalkowski, Z. W., *Aspiculuris tetraptera* Nietsch, 1821, in germfree and conventional mice, in *Germfree Research. Biological Effects of Gnotobiotic Environments. Proceedings of the IV International Symposium on Germfree Research*, Henghan, J. B., Ed., Academic Press, New York, 1973, 447.

13. Przyjalkowski, Z., Effect of intestinal flora and of a monoculture of *E. coli* on the development of intestinal and muscular *Trichinella spiralis* in mice, *Bull. Acad. Polon. Sci. Ser. Sci. Biol.*, 16, 433, 1968.

14. Przyjalkowski, Z., Starzynski, S., Pykalo, R., and Cabaj, W., Serological and pathological changes in germfree and conventional mice in single or mixed *Trichinella spiralis* and *T. pseudospiralis* infections, in *Recent Advances in Germfree Research. Proceeding of the VII International Symposium on Gnotobiology*, Sasaki, S. *et al.*, Eds., Tokai University Press, Tokyo, 1981, 447.

15. Przyjalkowski, Z., Cabaj, W., and Rykalo, R., Intestinal *Trichinella spiralis* and *Trichinella pseudospiralis* in germfree and conventional mice, *Prog. Fd. Nutr. Sci.*, 7, 117, 1983.

16. Przyjalkowski, Z. W., Intestinal microbiology and immune response of germfree and conventional mice in mixed trichinellosis, in *Germfree Research: Microflora Control and its Application to the Biomedical Sciences. Proceedings of the VIII International Symposium on Germfree Research*, Wostmann, B. S., Ed., Alan R. Liss, New York, 1985, 415.

17. Despommier, D. D., Antigens from *Trichinella spiralis* that induce a protective response in the mouse, *J. Immunol.*, 132, 898, 1984.

18. Cabaj, W., Cell transfer from germfree and conventional C3H donor mice which protect recipient mice from lethal infection with *Trichinella spiralis*, in *Gnotobiology and its Applications. Proceedings from the IX International Symposium on Gnotobiology*, Ed. Fond. Marcel Mérieux, Lyon, France, 1987, 242.

19. Wescott, R. B., Experimental *Nematospiroides dubius* infection in germfree and conventional mice, *Exp. Parasitol.*, 22, 245, 1968.

20. Chang, J. and Wescott, R. B., Infectivity, fecundity and survival of *Nematospiroides dubius* in gnotobiotic mice, *Exp. Parasitol.*, 32, 327, 1972.

21. Weinstein, P. P., Newton, W. L., Sawyer, T. K., and Sommerville, R. I., *Nematospiroides dubius*: Development and passage in the germfree mouse, and a comparative study of the free-living stages in germfree feces and conventional cultures, *Trans. Amer. Microsc. Soc.*, 88, 95, 1969.

22. Bezerra, M., Vieira, E. C., Pleasants, J. R., Nicoli, J. R., Coelho, P. M. Z., and Bambirra, E. A., The life cycle of *Schistosoma mansoni* under germfree conditions, *J. Parasitol.*, 71, 519, 1985.

23. Vieira, E. C., Bezerra, J. R., Nicoli, J. R., Silva, M. E., Mota-Santos, T. A., Coelho, P. M. Z., and Bambirra, E. A., Schistosomiasis mansoni in conventional and germfree mice, in *Gnotobiology and its Applications. Proceedings of the IXth International Symposium on Gnotobiology*, Ed. Fond. Marcel Mérieux, Lyon, France, 1987, 285.

24. Tavares, C. A., Cordiero, M. N., Mota-Santos, T. A., and Gazzinell, G., Artificially transformed schistosomula of *Schistosoma mansoni*. Mechanisms of acquisition of protection against antibody-mediated killing, *Parasitology*, 80, 95, 1980.

25. Resende, M. R., Silva, M. E., Silva, M. E., Silva, M. E., Bambira, E. A., Vieira, E. C., and Nicoli, J. R., The life cycle of *Trypanosoma cruzi* under axenic conditions, *J. Med.*, 23, 245, 1992.

26. Pedrosa, M. L., Nicoli, J. R., Silva, M. E., Silva, M. E., Vieira, L. Q., Bambirra, E. A., and Vieira, E. C., The effect of iron nutritional status on *Trypanosoma cruzi* infection in germfree and conventional mice, *Comp. Biochem. Physiol.*, 106A, 813, 1993.

27. Houser, B. B. and Burns, W. C., Experimental infection of gnotobiotic *Tenebrio molitor* and white rats with *Hymenolepis diminuta* (cestoda: cyclophyllidea), *J. Parasitol.*, 54, 69, 1968.

28. Newton, W. L., Weinstein, P. P., and Jones, J. F., A comparison of the development of some rat and mouse helminths in germfree and conventional guinea pigs, *Ann. N.Y. Acad. Sci.*, 78, 290, 1959.

Chapter IX

PATHOLOGY OVER THE LIFE SPAN OF THE GERMFREE RAT AND THE GERMFREE MOUSE

GENERAL ASPECTS

In the absence of a pathology of microbial origin we may expect a pathology relating to anomalies inherent to the GF state, to the special functional and/or metabolic status of the germfree animal, or to dietary inadequacies resulting from faulty diet sterilization. Typical anomalies are the enlarged ceca of the GF rodentia, occasionally leading to volvulus and death,[1,2] and the urinary calculi observed in GF rats fed the early casein-starch diets[3] (see Chapter III). In addition, the GF animal will demonstrate the degenerative pathology characteristic of aging, but now unobscured by superimposed bacterial insult. Cornwell and Thomas[4] describe typical cardiac fibrosis in old GF and CV male Lobund-Wistar rats (L-W rats), but find no effect of microbial status. They did find, however, reduced fibrosis in diet-restricted animals. Pathology of microbial origin, such as dental caries, is absent in the GF animal.[5,6] With the improvement of diets described in Chapter V, most of the pathological changes seen in the older GF animals now involve endocrine and endocrine-related glands and the liver.[7,8]

In earlier experiments in which casein-containing diets were used, a considerable number of older animals, both GF and CV, would show progressive degeneration of the tubular epithelium of the kidney. In the CV animals, on occasion this condition was aggravated by superimposed bacterial infection.[9] After natural ingredient diets such as L-485 (see Chapter V) were introduced, kidney lesions were uncommon and never severe enough to be the cause of death.[8] A more pronounced effect of the bacterial flora on kidney pathology was reported by Werder et al.[10] They found that whereas 83% of CV CFW$_{WD}$ mice would die within 12 months with cystic renal disease, their GF counterparts showed no renal pathology at that time. At 24 months all CV mice had died, 95% with cystic disease, but of the 65% of the GF animals that were now dead, only 4% showed cystic kidneys. An earlier study of GF and CV ICR mice by Walburg and Cosgrove[11] had already predicted these observations. The latter study had also confirmed the suspicion voiced by Pollard and Matsuzawa[12] that GF mice might harbor leukemogenic virus.

TUMORS AND TUMOR-RELATED VIRUSES

Autopsy data on older GF L-W rats showed the occurrence of solid tumors in liver, endocrine organs, breast, and lung. Pollard and Teah had been the first to report the occurrence of spontaneous tumors in GF rats.[13] In 1963 they reported the occurrence of tumors of spontaneous origin in GF Lobund-Wistar (L-W), but not in GF Fischer or GF Sprague-Dawley rats. Among several hundred older GF L-W rats autopsied they found 25 tumors, 18 adenofibromas involving the mammary gland, 3 involving the thymus, and 2 not clearly defined. Similar data were reported by Miyakawa et al.[14] At that time it would appear that except for the earlier-mentioned pathology, most of the lesions found in older GF rats were nonmalignant tumors.[15] However, a few years later Sacksteder et al.[16] reported several cases of mammary fibrosarcoma and one case of mammary adenocarcinoma in aging GF Fischer 344 rats, together with the occurrence of leukemia and isolated cases of renal and lung tumors. In a subsequent paper[17] covering ten generations of GF F344 male and female rats, Sacksteder found leukemia to be the most common neoplasm, followed by mammary tumors. Various other tumors, including testicular tumors, occurred in less than 10% of the animals. She mentions that although the overall occurrence of tumors was in the range observed in CV rats, significantly fewer solid tumors were observed in the GF than in the CV male rats. In 1979 Pollard and Luckert stated that:

> Liver tumors, ranging from benign nodules to carcinomas, developed spontaneously in 115 of 132 germfree L-W rats beyond the age of 30 months. Of these 11 were classified as hepatocellular carcinomas. In addition, the rats developed a high incidence of benign adenomas of endocrine glands and of the breasts.[18]

It thus became clear that as long as diet and housekeeping techniques allowed an optimal life span, both GF and CV rats and mice will eventually develop tumors.

At this point we have to consider the fact that the GF L-W rat is, by present testing methodology, virus free.[19] This is not assured for all strains of GF rats (see statement in Chapter I). GF mice, on the other hand, were found to harbor viruses.[20-22] Studies with LCM virus have demonstrated the transovarian/transplacental passage of this virus in gnotobiotic (GN) mice. At least seven strains of germfree mice were found to demonstrate leukemogenic virus after radiation.[12] In AKR mice the leukemogenic process even initiated spontaneously.[23] As a result the thymus glands were enlarged in the leukemic mice, while the spleens, livers, and kidneys were often swollen. Virus particles were observed in the cytoplasmic vacuoles of thymoma cells and in the intercellular spaces of thymus and spleen tissues. The thymic epithelial cells appeared to be a common reservoir for this type of tumor virus in the otherwise disease-free AKR mice. Mammary tumor virus was also demonstrated in GF C3H mice.[20] Obviously the virus factor has to be taken into account when considering aging-related pathology in GF rats and GF mice.

Thus far no virus has been reported in the commonly used strains of GF rats. This points to the GF rat as the more ideal model for the study of age-related tumors. Also, the *ad libitum*-fed GF rat will live, on the average, some 6 months longer than its CV counterpart (see Chapter II) thereby enabling, e.g., the development of liver tumors to an extent hardly seen in the CV rat.[18] That these tumors occur at all suggests the possible presence of carcinogens in the diet. Under CV conditions these could then be either activated or neutralized by the intestinal microflora,[24] thus adding another factor of uncertainty to the interpretation of the data. Pollard *et al.* conclude that: "In general, the oncogenic potentialities of germfree rodents resembled those of conventional counterpart animals. However, tumor-related changes were more clearly defined in the germfree rat."[25]

In the late 1960s the natural ingredient colony diet L-485[26] was introduced at the Lobund Laboratory. Diet L-485 consists mainly of soy and corn products, with added alfalfa and fortified with vitamins (see Chapter V). The absence of casein and other dairy proteins from the diet almost totally prevented the aforementioned kidney pathology, and increased the average life span of both the GF and the CV male L-W rats by several months. Since a fair number of GF males and females now lived beyond the age of 30 months, a high percentage was found to develop neoplastic lesions of the liver in addition to tumors of endocrine organs, uterus, and lung (Table 1).[18] While CV rats also showed an increased life span when maintained on diet L-485, only relatively few managed to survive beyond 30 months of age. However, comparing male CV rats aged 25 to 29 months with a comparable GF group, the impression is gained that the occurrence of liver tumors might have been of the same magnitude. Both GF and CV rats showed gamma glutamyl transpeptidase foci in the liver, indicating preneoplastic and/or neoplastic lesions, as early as the first year of life.[27] This led to a suspicion, although never verified, that diet L-485 might contain potentially oncogenic material.

At the Symposium of the Effects of Dietary Restriction on Aging (Notre Dame, 1988) the pathology of aging was reviewed. Extensive details of pituitary tumor formation were given by Kovacs *et al.*[28] They found prolactin-producing adenomas uncommon, but gonadotropic adenomas containing FSH and LH occurring frequently in older animals, though less in the GF than in the CV rats. Pollard and Luckert[7] reviewed the

TABLE 1

Spontaneous Tumors Developed by Germfree
Lobond-Wistar Rats Over 30 Months of Age

	90 Male (%)	42 Female (%)
Liver tumors	76 (84.4)	39 (92.8)
Adrenal medulla adenoma	72 (80.0)	20 (47.6)
Pituitary adenoma	48 (53.0)	32 (76.0)
Thymoma	41 (45.5)	28 (66.6)
Breast adenofibroma	26 (28.8)	29 (69.0)
Uterine fibroma	—	18 (42.8)
Lung adenoma	12 (13.3)	9 (21.4)
Thyroid adenoma	12 (13.3)	6 (14.2)
Parathyroid adenoma	25 (27.7)	10 (23.8)
Adrenal cortex adenoma	18 (20.0)	8 (19.0)
Prostate adenocarcinoma	9 (10.0)	

Table taken from Pollard, M. and Luckert, P.H., *Lab. Anim. Sci.,*
29, 74, 1979. With permission.

occurrence of liver tumors, adrenal medullary tumors (which they later classified as hyperplasia), and tumors of other endocrine organs. In addition, extensive data were given on the spontaneous occurrence of prostate adenocarcinoma in both GF and CV male L-W rats older than 2 years. This syndrome had been described earlier and is apparently unique for the L-W strain of rats.[29-31] The condition is described as a moderately differentiated adenocarcinoma comparable to the human disease. In about 70% of the tumor-bearing rats metastasis occurs primarily via the abdominal lymphatic channels to the lungs, but also by extensions through the capsule of the organ in the form of small round solid tumors (pearls) attached to peritoneal surfaces, including surfaces of the visceral organs (Figures 1A, B, and C). This condition is promoted by testosterone and by high dietary fat.

In general, the presently available data seem to suggest that, other conditions being comparable, spontaneous tumors occur somewhat less frequently in GF than in CV rats. The same impression is gained from a number of studies with direct-acting carcinogens. Laquer *et al.*[32] administered methylazoxymethanol or its acetate by various routes to GF and CV Sprague-Dawley rats, resulting in tumors in the intestine, kidney, and liver. They found half as many intestinal tumors in the GF as in the CV rats. Later, Reddy *et al.*[33] in a study of colon carcinogenesis, reported that upon administration of N-methyl-N'-nitro-N-nitrosoguanidine to weanling CD Fischer rats, fewer tumors were found in the GF than in the CV animals. The authors point out, however, that conditions in the GF gut differ substantially from those in the CV gut: the average age of the enterocytes is greater, and bile acids are in a higher concentration in the GF gut and are in the primary, unconjugated form. Also the local oxidation-reduction potential is about 300 mV more positive (see Chapter III) — all factors which could affect tumorigenesis. Sasajima *et al.*,[34] who with

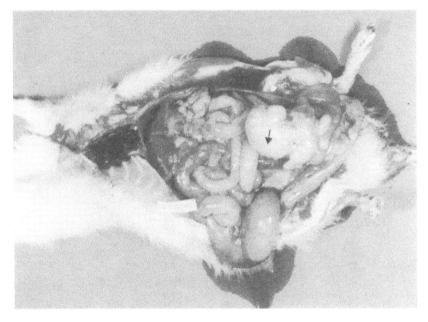

FIGURE 1A
Spontaneous prostate adenocarcinoma of a 36-month-old germfree Lobund-Wistar rat (arrow; weight 20 g). (Courtesy of Dr. Morris Pollard.)

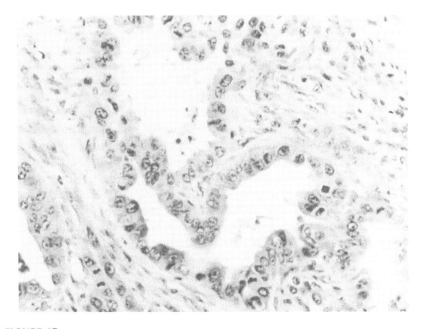

FIGURE 1B
Histopathology of a spontaneous adenocarcinoma of the prostate of a germfree Lobund-Wistar rat.

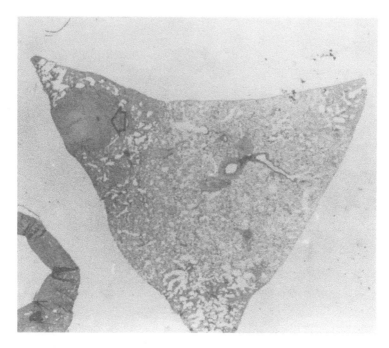

FIGURE 1C
Histopathology of metastatic prostate adenocarcinoma (arrow) in the lung of a germfree
Lobund-Wistar rat.

oral administration of N-methyl-N-amylnitrosamine were able to produce
tumors in the esophagus of GF and CV male Wistar rats within 15 weeks,
again found that with their experimental protocol carcinoma developed
in 60% of the CV, but only in 14% of the GF rats. However, the possibility
of bacterial modification with enhanced potential of the above carcinogens
by the CV animals must be taken into account here (see Chapter XI).

Reviewing the final autopsy reports of the aforementioned Lobund
Aging Study, the data do not indicate any really significant difference in
the rate of occurrence of spontaneous tumors except in the case of the
pancreas, where no tumors were found in the GF rats. However, this
evaluation is essentially flawed since the average time of death of the
autopsied GF rats was 33 months, against 27 months for the CV rats. For
the adrenal, parathyroid, and mammary tumors the data seem to suggest
a lower occurrence in the GF rats (p values 0.17, 0.18, and 0.16, respectively) which, together with the lower average age of the CV animals at
autopsy, intimates that these endocrine-related tumors will develop more
slowly in GF than in CV rats. A more direct comparison is available for
the spontaneous occurrence of the aforementioned prostate tumors. At
the age of 26 months only 5% of GF rats had developed prostate cancer,
against 26% in the CV animals.[35] These data, and some of the following
observations on mice, point to the fact that for reasons not quite clear

endocrine-related tumors develop slower in GF than in CV rats and mice. One might speculate that this could be related to the early lower energy intake and the persistently lower oxidative exposure of these GF animals (see Chapter IV), especially in view of the fact that in the GF rat a forced reduction of dietary intake by about 30% will drastically reduce the occurrence of prostate tumors.[7] Interestingly, this reduction in dietary intake tends to increase the level of circulating testosterone.[36]

Although the apparently ubiquitous presence of virus in GF mice makes for a somewhat "unclean" model, data on the occurrence of tumors confirm, to an extent, the results obtained with GF rats. As mentioned earlier, in 1965 Kajima and Pollard[20] reported lymphoid leukemia in AKR mice. They also found mammary adenocarcinomas in two GF C3H mice. In both instances the thymus and the mammary tumor revealed B type virus-like particles. A few years later Pilgrim and Labrecque reported that the incidence of mammary tumors in GF and CV C3H mice was virtually identical.[37] At about the same time Walburg and Cosgrove[11] reported leukemia and tumors of the lung, breast, bone, adrenal, gonads, and liver in aging GF ICR mice. No remarkable difference in the occurrence of solid tumors was seen between GF and CV animals. Seibert et al.[38] studied the Hodgkin's-like syndrome which develops in GF and CV SJL/J mice and is presumed to have a virus as a causative factor. After 4 to 5 months of age the animal's immune potential starts to decline, and about 3 months later the syndrome becomes obvious. Both GF and CV mice showed similar impairment of humoral immunity associated with increasing age and also a comparable extent of pathology, but in the CV mice the condition appeared aggravated by secondary infection (see also Chapter VII).

On the other hand, Mizutani and Mitsuoka[39] report a much lower occurrence of liver tumors in irradiated (5 Mrad) 48-week-old GF than in CV C3H/He mice (39 vs. 82%). They suggested that microbial conversion of procarcinogens in the diet may be at least one of the reasons for the observed difference. However, when the same authors studied spontaneous polyposis in GF and CV BALB/c mice after an incidental finding in GF mice, it was found that in 1-year-old animals the GF mice showed about twice as many polyps as their CV counterparts (79 vs. 44%).[40] They explained this apparently beneficial action of the microflora by a possible activation of cytotoxic cells under CV conditions.

CONCLUSIONS

All in all, the data thus far suggest the following:

1. Upon aging, GF and CV rats develop similar tumors. The same appears true for GF and CV mice.
2. The development of these tumors appears to be slower in GF than in CV rats, conceivably related to the GF animals' lower dietary intake

and O_2 consumption at a younger age (see Chapter IV, Energy Metabolism). There are indications that this may also be true for GF mice.

3. This, and the occurrence of early precancerous conditions in the liver of both GF and CV rats may point to the presence of potential carcinogens in the diet and/or bedding, possibly activated (or inactivated?) by members of the intestinal microflora.[41]

The ultimate answer will have to come from GF mice and GF rats fed only a chemically defined diet and kept on pure cellulose bedding (see Chapter VI).

REFERENCES

1. Wostmann, B. S. and Bruckner-Kardoss, E., Development of cecal distention in germfree baby rats, *Am. J. Physiol.*, 197, 1345, 1959.
2. Djurickovic, S. M., Ediger, R. D., and Hong, C. C., Volvulus at the ileocecal junction of germfree mice, *Lab. Anim.*, 12, 219, 1978.
3. Gustafsson, B. E. and Norman, A., Urinary calculi in germfree rats, *J. Exp. Med.*, 116, 273, 1963.
4. Cornwell, G. G. and Thomas, B. P., Cardiac fibrosis in the aged germfree and conventional Lobund-Wistar rat, in *Dietary Restriction and Aging*, Snyder, D. L., Ed., Alan R. Liss, New York; *Prog. Clin. Biol. Res.*, 287, 69, 1989.
5. Orland, F. J., Blayney, J. R., Harrison, R. W., Reyniers, J. A., Trexler, P. C., Wagner, M., Gordon, H. A., and Luckey, T. D., The use of the germfree animal technique in the study of experimental dental caries, *J. Dent. Res.*, 33, 147, 1954.
6. Fitzgerald, J., Dental caries research in gnotobiotic animals, *Caries Res.*, 2, 139, 1968.
7. Pollard, M. and Luckert, P. H., Spontaneous diseases in aging Lobund-Wistar rats, *Prog. Clin. Biol. Res.*, 287, 51, 1989.
8. Snyder, D. L., Pollard, M., Wostmann, B. S., and Luckert, P., Life span, morphology, and pathology of diet-restricted germfree and conventional Lobund-Wistar rats, *J. Gerontol.*, 45, B52-58, 1990.
9. Gordon, H. A., Bruckner-Kardoss, E., and Wostmann, B. S., Aging in germfree mice. Life tables and lesions observed at natural death, *J. Gerontol.*, 21, 380, 1966.
10. Werder, A. A., Amos, M. A., Nielsen, A. H., and Wolfe, G. H., Comparative effects of germfree and ambient environments on the development of cystic kidney disease in CFW_{WD} mice, *J. Lab. Clin. Med.*, 103, 399, 1984.
11. Walburg, H. E. and Cosgrove, G. E., Aging in irradiated and unirradiated germfree ICR mice, *Exp. Gerontol.*, 2, 143, 1967.
12. Pollard, M. and Matsuzawa, T., Radiation-induced leukemia in germfree mice, *Proc. Soc. Exp. Biol. Med.*, 116, 967, 1964.
13. Pollard, M. and Teah, B. A., Spontaneous tumors in germfree rats, *J. Natl. Cancer Inst.*, 31, 457, 1963.
14. Miyakawa, M., Sumi, Y., and Uno, Y., Spontaneous tumors in Japan-born germfree rats, in *Germfree Research. Biological Effects of Gnotobiotic Environments. Proceedings of the IV International Symposium on Germfree Research*, Heneghan, J. B., Ed., Academic Press, New York, 1973, 123.
15. Pollard, M., Senescence in germfree rats, *Gerontologia*, 17, 333, 1971.

16. Sacksteder, M. R., Kasza, L., Palmer, J. L., and Warren, J., Cell transformation in germfree Fischer rats, in *Germfree Research. Biological Effects of Gnotobiotic Environments. Proceedings of the IV International Symposium on Germfree Research,* Heneghan, J. B., Ed., Academic Press, New York, 1973, 153.
17. Sacksteder, M. R., The occurrence of spontaneous tumors in the germfree F344 rat, *J. Natl. Cancer Inst.,* 57, 1371, 1976.
18. Pollard, M. and Luckert, P. H., Spontaneuos liver tumors in aged germfree Wistar rats, *Lab. Anim. Sci.,* 29, 74, 1979.
19. Pollard, M. and Kajima, M., Leukemia in germfree rats, *Proc. Soc. Exp. Biol. Med.,* 121, 585, 1966.
20. Kajima, M. and Pollard, M., Detection of viruslike particles in germfree mice, *J. Bacteriol.,* 90, 1448, 1965.
21. Walburg, H. E., Upton, A. C., Tyndall, R. L., Harris, W. W., and Cosgrove, G. E., Preliminary observations on spontaneous and radiation-induced leukemia in germfree mice, *Proc. Soc. Exp. Biol. Med.,* 118, 11, 1965.
22. Kajima, M. and Pollard, M., Distribution of tumor viruses in germfree rodents, in *Sixth International Congress for Electron Microscopy,* Vol. II, Uyeda, R., Ed., Maruzen Co. Ltd., Tokyo, 1966, 215.
23. Pollard, M., Kajima, M., and Teah, B. A., Spontaneous leukemia in germfree AK mice, *Proc. Soc. Exp. Biol. Med.,* 120, 72, 1965.
24. Weisburger, J. H., Colon carcinogens: their metabolism and mode of action, *Cancer,* 28, 60, 1971.
25. Pollard, M., Kajima, M., and Lorans, G., Tissue changes in germfree rats with primary tumors, *J. Reticuloend. Soc.,* 5, 147, 1968.
26. Kellogg, T. F. and Wostmann, B. S., Stock diet for colony production of germfree rats and mice, *Lab. Animal Care,* 19, 812, 1969.
27. Pollard, M., Luckert, P. H., and Adams, R. A., Detection of gamma glutamyl transpeptidase in the liver of germfree and conventional rats, *Lab. Anim. Sci.,* 32, 147, 1982.
28. Kovacs, K., Ryan, N., Sano, T. Stefaneanu, L., Ilse, G., and Asa, S. L., Adenohypophysical changes in conventional, germfree and food-restricted aging Lobund-Wistar rats. A histologic, immunocytochemical and electron microscopic study, *Prog. Clin. Biol. Res.,* 287, 181, 1989.
29. Pollard, M., Spontaneous prostate adenocarcinomas in aged germfree Wistar rats, *J. Natl. Cancer Inst.,* 51, 1235, 1973.
30. Pollard, M., Animal model of human disease. Metastatic adenocarcinoma of the prostate gland, *Am. J. Pathol.,* 86, 277, 1977.
31. Pollard, M., Luckert, P. H., and Snyder, D., Prevention of prostate cancer and liver tumors in L-W rats by moderate dietary restriction, *Cancer,* 64, 686, 1989.
32. Laquer, G. L., McDaniel, E. G., and Matsumoto, H., Tumor induction in germfree rats with methylazoxymethanol (MAM) and synthetic MAM acetate, *J. Natl. Cancer Inst.,* 39, 355, 1967.
33. Reddy, B. S., Weisburger, J. H., Narisawa, T., and Wynder, E. L., Colon carcinogenesis in germfree rats with 1,2-dimethylhydrazine and N-methyl-N'-nitro-N-nitrosoguanidine, *Cancer Res.,* 34, 2368, 1974.
34. Sasajima, K., Taniguchi, Y., Nakao, M., Tamura, H., Adachi, K., and Shirota, A., Sequential studies on esophageal carcinoma induced in conventional and germfree rats by N-methyl-N-amylnitrosamine, in *Recent Advances in Germfree Research, Proceedings of the VII International Symposium on Gnotobiology,* Sasaki, S. *et al.,* Eds., Tokai University Press, Tokyo, 1981, 657.
35. Luckert, P. H., Snyder, D., and Pollard, M., Prevention of spontaneous prostate cancer in germfree Lobund-Wistar rats relative to body and organ weights, in *Experimental and Clinical Gnotobiology. Proceedings of the 10th International Symposium on Gnotobiology,* Heidt, P. J. *et al.,* Eds., *Microecol. Ther.,* 20, 401, 1990.

36. Wostmann, B. S., Snyder, D. L., and Johnson, M. H., Metabolic effects of the germfree state in adult diet-restricted male rats, in *Experimental and Clinical Gnotobiology. Proceedings of the Xth International Symposium on Gnotobiology,* Heidt, P. J. *et al.,* Eds., *Microecol. Ther.,* 20, 383, 1990.

37. Pilgrim, H. I. and Labrecque, A. D., The incidence of mammary tumors in germfree C3H mice, *Cancer Res.,* 27, 584, 1967.

38. Seibert, K., Pollard, M., and Nordin, A., Some aspects of humoral immunity in germfree and conventional SJL/J mice in relation to age and pathology, *Cancer Res.,* 34, 1707, 1974.

39. Mizutani, T. and Mitsuoka, T., Effect of intestinal bacteria on the incidence of liver tumors in gnotobiotic C3H/He mice, *J. Natl. Cancer Inst.,* 63, 1365, 1979.

40. Mizutani, T. and Mitsuoka, T., Spontaneous polyposis in the small intestine of germfree and conventionalized BALB/c mice, in *Germfree Research: Microflora Control and Its Application to the Biomedical Sciences. Proceedings of the VIII International Symposium on Germfree Research,* Wostmann, B. S., Ed., Alan R. Liss, New York, 1985, 297.

41. Pollard, M., Development of model systems for cancer research, in *Recent Advances in Germfree Research. Proceedings of the VII International Symposium on Gnotobiology,* Sasaki, S. *et al.,* Eds., Tokai University Press, Tokyo, 1981, 621.

Chapter X

APPLICATIONS: PAST, PRESENT, AND FUTURE. PART I

INTRODUCTION

As mentioned in Chapter I, the GF and the GN animal were developed to have available an animal model of the highest possible definition. This furthermore necessitated the use of inbred strains, and eventually led to the development of the chemically defined "antigen-free" diets (see Chapter VI). The practical application of this model actually started as soon as sufficient numbers of animals could be produced to conduct meaningful experiments. The following chapters will list applications characteristic of the potential of the gnotobiote. This survey will not and cannot be all inclusive, as it only tries to point out the specific potential of the animal model, either GF or harboring a defined microbial ecosystem.

Colony production of GF rats had taken off only in the late 1940s (see Chapter I); but already in 1951 du Vigneau *et al.* had published their study, The Synthesis of "Biologically Labile" Methyl Groups in the Germfree Rat,[1] in which they showed that GF rats fed deuterium oxide would incorporate deuterium in the methyl groups of compounds like choline and creatine. A few years later, collaboration between the Walter G. Zoller Memorial Dental Clinic of the University of Chicago and the Lobund Laboratory of the University of Notre Dame confirmed what many had expected: that in the absence of a microbial flora rats do not develop dental caries.[2]

Thus, the GF and the GN animal obviously are tools — animal models which make it possible to study biological and pathological events under rigorously controlled conditions. This is especially true of the GF rat, which appears to be "germfree" in the truest sense of the definition given by Wagner:[3] being free not only of bacteria, but also of viruses[4] and any other contaminants (see Chapter I for the restrictions placed on this statement). Following are applications in the biomedical and veterinary field that are illustrative of the potential applications of this animal model.

MICROBIAL ACTION AND INTERACTION IN GNOTOBIOTIC SYSTEMS: COLONIZATION RESISTANCE AND TRANSLOCATION STUDIES

Controlled Microbial Association

In 1965 Dymsza et al.[5] reported the establishment of a GN microflora consisting of four species "of predominating microorganisms of human intestinal origin" isolated from young males, in young GF rats. Weight gain during the 3-week experimental period was comparable to that of GF controls, and cecal size was not reduced. Actually, this study followed an earlier one by Gibbons et al.[6] who had been able to establish a mixture of organisms indigenous to the human gingival area in GF mice. Also in 1965, Schaedler et al.,[7] working in Dubos' laboratory at Rockefeller University, reported the establishment of a stable "hexaflora" in ex-GF mice, an ecosystem comprising mouse intestine-derived *Lactobacillus brevis, Enterococcus faecalis, Staphylococcus epidermidis, Bacteroides fragilis* var. *vulgatus, Enterobacter aerogenes,* and a *Fusibacterium* sp. With slight variation and/or additions, this GN mouse would become a model for the study of biomedical phenomena for which the presence of a defined and stable microflora of indigenous species was desirable.

Looking over the multitude of studies reported during the last 30 years that could be categorized under the above heading, it would seem that they mostly fall into several categories:

1. Study of the capacity of specific microorganisms to establish as mono- or polyassociates in GF animals.
2. Study of the effect of specific microorganisms on function and metabolism of the host, as monoassociates or within an otherwise defined microbial ecosystem.
3. Study of the interaction of specific microbial species and/or strains under controlled conditions.
4. Study of bacteria of human and other origin under GN conditions.
5. Study of the interaction of yeasts and bacteria.

Monoassociation

The purposely monoassociated animal is a model that makes it possible to study aspects of the effect of the associate on the host. It should be kept in mind, however, that in the absence of other microorganisms these effects may be greatly altered, be it only because the host may now be confronted with much higher numbers of that organism, presumably creating unusual intestinal conditions. Still, some vital information may be garnered from such studies, as is illustrated by the work on dental caries and periodontal disease mentioned earlier.

In 1964 Margard and Peters[8] had associated albino ND-2 mice (derived from the Lobund Laboratory) maintained on Ralston Purina diet 5010C, with a mouse-derived strain of *Salmonella typhimurium*. Although several of the mice died during the postinoculation period, the usual clinical symptoms were not observed and most of the mice survived and later reproduced. When Wostmann *et al.*[9,10] associated GF rats with *S. typhimurium* a similar situation was found. Weight loss was observed, and again some of the animals died. The others fully recovered with no signs of impaired health. However, in these studies the effect of diet became obvious. Rats maintained on diets 5010C and L-485 (see Chapter V, Table 2) showed moderate weight loss and no death, whereas those fed the more defined diet L-462 (based on grains and milk protein, see Chapter V, Table 1) showed much more severe weight loss, while about 1/3 of the animals died. Thus, diet may be very important in the interpretation of these studies. In this case one could speculate that diets 5010C and L-485, after autoclaving, were more immunogenic than diet L-462, resulting in a more "prepared" immune system in the originally GF rats.

Monoassociation has been used by Sasaki and co-workers[11] to study performance of a number of microorganisms in CD-1 mice as a basis for the formation of an "artificial" but defined microflora. Included were *Escherichia coli, Streptococcus faecalis, Lactobacillus acidophilus, Clostridium perfringens, Pseudomonas aeroginosa, Candida albicans, Shigella flexneri, Vibrio cholerae, El Tor vibrio,* and *Vibrio parahemolyticus.* The mice were maintained on a Japanese analogue of the Purina 5010C autoclaved diet. All organisms became established and most persisted, although lactobacillus, pseudomonas, and candida had to be administered in large numbers to maintain persistency. None of these monoassociates caused infectious death, though they persisted for a long time in the gut. A later study of *V. cholerae* V86 El Tor Inaba by Sack and Miller[12] and by Shinamura[13] showed that this organism readily colonizes the intestinal tract of GF mice and induces specific IgA antibodies to vibrios. However, in spite of this the animals remained healthy. A later study by Shinamur et al.[14] showed that when the GF mice were orally infected at the age of only 3 days with this especially toxinogenic strain, 2/3 of the animals died, most within 4 days. Infected 4 to 6 days later, about 30% of the animals died, but at 14 days no deaths occurred. Thus, when this vibrio was administered at a very early age, immune protection did not develop soon enough and a substantial number of the associated mice died within a few days.

Fox *et al.*[15] have used mice monoassociated with *Helicobacter felis* to study chronic gastritis, now considered largely of bacterial origin. Both GF and CV mice developed an anti-*H. felis* IgM antibody response during the 8 weeks after infection. Thereafter significant levels of IgG response were seen, which remained elevated throughout the 50-week course of the study.

Microbial Interaction

The early studies of the controlled interaction of microorganisms in GN rats and GN mice have mostly been carried out in France, with Sacquet, Ducluzeau, and/or Raibaud as senior investigators. In 1966 this group had reported on the establishment of up to 58 defined microbial species isolated from CV rats into ex-GF rats. They mention specifically that under their experimental conditions E. coli and S. pyogenes tended to dominate in the intestinal tracts of the gnotobiotes.[16] The latter observation presumably gave rise to their studies of the interaction of these bacteria in the GN mouse. They found that while established as monoassociates they each reached approximately the same numbers in the gut; as diassociates E. coli E_{12} would depress the establishment of S. pyogenes by a factor of 100. Also, the prior establishment of S. pyogenes did not interfere with the later establishment of E. coli, but E. coli when first introduced, would retard establishment of S. pyogenes by at least one week.[17] They subsequently vaccinated mice by i.p. injection of killed bacteria from either strain. In GF mice treated with E. coli the growth of subsequently administered live bacteria was retarded for several hours, but after 24 h the numbers of E. coli in vaccinated and control animals proved to be the same. Vaccination with S. pyogenes did not retard the subsequent establishment of p.o. administered E. coli, but vaccination with E. coli did retard the subsequent establishment of S. pyogenes. The establishment of the above strains in the intestine proved to be less efficient in producing specific antibodies than the i.p. injection of the killed material.[18]

Along similar lines, these workers reported two different studies on the mutual interaction of two different strains of E. coli[19,20] and of different strains of staphylococci[21] established in the GN mouse. A few years later workers in Japan studied the effect of lactobacilli in controlling the bacterial composition of other bacterial species in the digestive tract of polyassociated gnotobiotic rats. Among other things, they found that lactobacilli colonized preferentially in the stomach, preventing staphylococci from establishing there.[22] However, this potential of lactobacilli may depend on the local nutritional conditions and the bacteria encountered. Earlier, Ducluzeau et al.[23] had reported that when GF mice were diassociated with a strain of Lactobacillus and a Ristella sp., the two coexist without interference. When, however, sufficient lactose was added to the diet, Ristella was eliminated. In a more complex polyassociated situation, when E. coli, C. bifermentans, and S. pyogenes were also present, again only Ristella was eliminated by lactose addition. No signs of a shift in pH due to lactic acid production was established. Recently, Nielsen et al.[24] again studied the possible interaction of two E. coli strains (BJ18, rifampicin-resistant, and BJ19, nalidixic acid-resistant) after introduction into GF rats. Although the colonization ability of the lumen, mucus, and epithelium of the cecum were obviously different, the relative distribution of the organisms over these niches was the same. Even when introduced into CV rats, where colonization resistance strongly reduced their numbers, the relative distribution of the organisms appeared to be comparable.

Some time later the use of the gnotobiote in the study of more complicated microbial ecosystems of human origin was explored. Advocated by the French group,[25] it was found that after inoculation with cecal or fecal material, or with a large number of specific organisms of human origin, the recipient gnotobiote mirrored the original inoculum quite well — well enough to enable the study of a specific aspect of the ensuing intestinal ecosystem under more controlled conditions.[26] Gismondo *et al.*[27] confirmed this, stating that colonization of the GF mouse intestine was achieved by most of the human fecal bacteria that had been inoculated, and that these microfloras were stable with time. Based on these observations, Corthier *et al.*[28] studied the interaction of the cytotoxigenic, human-derived *Clostridium difficile* VPI 10463 in GN mice earlier monoassociated with a number of human-derived organisms. Protection occurred in mice previously inoculated with *E. coli* or *Bifidobacterium bifidum*. Intestinal cytotoxin production was greatly reduced in the surviving mice, although *C. difficile* numbers were not reduced to a great extent.

To further study the various factors involved in the persistence of *V. cholerae* in the intestinal tract, researchers again turned to the GN mouse model. Miller and Feeley[29] administered a culture of *V. cholerae* Ogawa 395 into the esophagus of GF mice, and thereafter similarly administered cultures of bacteroides, clostridia, lactobacilli, proteus, pseudomonas, and streptococci alone or in various combinations. No single strain or species was found that by itself would eliminate *V. cholerae* from the intestine within 14 days. Only a combination of *E. coli, P. mirabilis,* and *S. faecalis* was eventually able to eliminate *V. cholerae* within that time span.

Obviously the gnotobiotic model is an ideal system to study conjugal transfer of plasmid DNA, since the "interested parties" and the possibly affecting microorganisms can be strictly controlled. Schlundt *et al.*[30] have studied the transfer of plasmid $_p$AMβ1 between two strains of *L. lactis* subsp. *lactis* in the intestinal tract of GN rats. They found a considerable number of transconjugants throughout the intestine, and noted that plasmid $_p$AMβ1 seemed to have conveyed to the recipient a competitive advantage in the small, but not in the large intestine. Duval-Iflah *et al.*[31] investigated the role of helper elements in the mobilization of $_p$BR recombinant plasmids from genetically engineerd ori⁻ *E. coli* K12 to other K12 strains and to wild-type *E. coli* of human origin *in vitro* and in GN mice. Their results indicated that over 50% of the wild-type strains were able to promote transfer of pBR *oriT*⁻ plasmids *in vitro*, but not in GN mice.

Bacteria-Yeast Interaction

The incidence of human disease caused by *Candida albicans* has increased steadily over the years, pointing to the need for a reliable model for candidiasis that mimics the systemic disease in humans, and could suggest means to control its pathogenicity. In 1966 Phillips and Balish[32] reported the establishment of *C. albicans* in both GF and CV Swiss-Webster mice, but their results could not be confirmed by Nishikawa *et al.*,[33] who could establish the organism only in GF ICR mice but not in the SPF mice used in the study. When *E. coli* was subsequently established in the *C. candida*-associated ICR mice, *C. albicans* levels were significantly depressed, whereas it proved impossible to establish *C. albicans* in *E. coli*-monoassociated mice. Immunoelectrophoresis of the serum of the *C. albicans*-associated mice in this study suggested a certain level of antibody formation. Obviously the levels of other microorganisms, notably *E. coli*, in the CV mice may have played a role here. Differences in strain and methodology may have added to the discrepancy.

A few years later, Rogers and Balish[34] detailed the infection of CV Swiss-Webster mice with *C. albicans* strain B311 and found that the organism established preferentially in the kidney for up to 24 days. Thereafter, slow elimination started. They also reported the establishment of *C. albicans* in GF rats. In this case *C. albicans* was followed in the kidney, where it reached peak levels at day 10, to be eliminated around day 24. Recently Balish and co-workers[35] studied the immune response to *C. albicans* in GF athymic (nu/nu) mice and euthymic (nu/+) mice (BALB/c background). This comparison appeared to indicate the formation of protective antibody, as was suggested earlier by the Nishikawa group.[33] The authors suggest that T cell-mediated immunity may also have given a certain amount of protection, but did not play a prominent role. In a following study this group established *C. albicans* in GF SCID mice.[36] Large, viable populations were found in all sections of the intestinal tract, but only superficial mucosal candidiasis and no progressive disseminated candidiasis were evident. Colonization of the gut with a bacterial microflora enhanced the resistance of these mice to disseminated candidiasis. Apparently, innate immune mechanisms (phagocytic and/or NK cells), in the absence of functioning T and B cells, play an important role in the resistance of the SCID mouse to mucosal and disseminated candidiasis.

Colonization Resistance

Some of the above observations had suggested the possibility that intestinal microbial populations, depending on their composition, might be able to prevent other bacterial species from establishing. This was first explored by Raibaud *et al.*[37] who observed that although both spores and vegetative cells of *C. perfringens* established readily in GF mice, both were rapidly eliminated from the gut of CV mice, the spores largely as vegetative cells. This phenomenon was soon to become known as "colonization resistance" or CR.[38,39] CR has been explored extensively because of its importance in the treatment of medical conditions where the establishment of especially Gram-negative pathogens could create a major danger to the patient.

Koopman et al.[40] had associated GF mice with a number of specific bacterial strains isolated from CV mice in an effort to produce a "normalizing flora" which would bring specific GF characteristics (cecal enlargement, bile acid spectrum) back within the CV range. They also considered the potential for CR of these floras, and concluded that CR potential might correlate with propionic acid production, although "this could not be the sole mechanism for CR." In the same year, Hazenberg et al. reported having associated GF mice with a human intestinal microflora, which established well and remained stable and comparable to the original inoculum. This human-derived flora reduced the cecum to CV size and induced CR against P. aeruginosa. More recently, Raibaud et al.[41] inoculated GF mice with a human flora and again observed that the resulting mouse flora resembled the human flora, with the exception of Bifidobacterium sp. and lactobacilli, which possibly got lost during transfer. In a follow-up of these studies the French group investigated the effect of bran in the human diet on the CR of the human intestinal floras transferred into mice. As before, the resulting mouse microfloras resembled the original human floras quite well. They found absolute barrier effects against C. perfringens and S. aureus, independent of any addition of bran to the human diet. CR against P. aeruginosa and C. difficile, however, seemed to depend on dietary variation.

Whereas the importance of the study of the CR potential of the human intestinal microflora is obvious, not so obvious is the use of microfloras derived from the golden hamster. Work by Lusk et al.[42] indicated that this animal's microflora had an inherent resistance to C. difficile. This resistance could be broken by microflora alterations following the administration of an antibiotic, in this case Clindamycin. To obtain a model in which this phenomenon could be studied, Su et al.[43] associated GF mice with a hamster flora by inoculating the mice p.o. with hamster cecal contents. They report that the flora which established in the ex-GF mice appeared to closely resemble the hamster flora, and that it had retained the barrier effect against C. difficile. Treatment with erythromycin removed E. coli, but maintained the dominant anaerobic flora, which retained its barrier effect. More recently, Wilson and Freter[44] cultured the cecal flora of hamsters in a continuous-flow system which had first been inoculated with C. difficile and E. coli. They also introduced hamster flora cultures in GN mice that had been associated with these organisms. In both cases the numbers of clostridia and coli bacteria were dramatically reduced. Introduction of the culture of the hamster microflora also reduced the cecum of the GN mice to CV size. They then proceeded with the introduction of large numbers of specific isolates of the hamster flora in the GN mice and found significant suppression of C. difficile and E. coli, depending on the isolates.

Guiot[45] had followed up on the aforementioned idea that volatile fatty acids production by elements of the intestinal microflora might be the mechanism responsible for CR, but his experiments with CV rats could not confirm this. Su and co-workers,[46] using a different approach, associated GF mice with different hamster floras. Although the hamster-derived microorganisms produced substantial amounts of acetic, propionic, and butyric acid, *in vitro* studies indicated that at those concentrations the growth of *C. difficile* was not inhibited. They concluded that volatile fatty acids alone would not inhibit intestinal colonization by *C. difficile*, and that other, as yet unknown factors were involved in the CR exhibited by the hamster-derived microfloras.

In a further effort to understand the etiology of CR, French workers concentrated on the antagonism against *C. perfringens*. In 1987 Yurdusev *et al.*[47] reported that they had achieved this with three microorganisms: *Bacteroides thetaiotaomicron, Fusibacterium necrogenes,* and a *Clostridium* sp. Similar results were reported by Boureau *et al.*[48] who associated mice with *C. indolis, C. cocleatum,* and *Eubacterium* sp. In a later publication they describe details of the mucosal colonization of these organisms, and speculate that competition for nutrients may be at the base of the CR phenomenon.[49] Subsequently, Yurdusev *et al.* reported that only two bacteria, *Bacteroides thetaiotaomicron* and *Fusobacterium necrogenes*, this time derived from the feces of piglets, were sufficient to keep *C. perfringens* in check. Their conclusion: "… a reversible bacteriostasis induced by the inhibitory strains acting together continuously, and hindering the target strain from utilizing available nutrients, was responsible for this antagonism."[50] Thus, a competition for specific nutrients required by the "resisted" organism appears to be at this moment one of the most plausible causes for the phenomenon labeled "colony resistance". Much earlier, Freter *et al.*[51] had already come to a similar, albeit more general conclusion.

As a variant of the work on CR the studies of Wells *et al.*[52] with *E. coli chi* 1776 should be mentioned. The original purpose of this artificially engineered organism had been to have a strain of coli bacteria available that could be used for experimentation, but would not be able to survive outside the laboratory. A debilitated strain of *E. coli* K-12 had been developed for the use in recombinant DNA research, designated *E. coli chi* 1776, which not only had special growth requirements, but also showed increased sensitivity to UV irradiation, bile acids, detergents, antibiotics, and temperature.[53] This organism was not supposed to establish in CV animals, but it was considered that a better test would be to try to establish the organism in GF rodents, where no indigenous microbial competition would exist and the immune system would be relatively inactive. Although a similar debilitated coli organism had shown some measure of survival in CV rodents, *E. coli chi* 1776 could not be established even in GF rodents. This presumably was because of its sensitivity to the GF rodent's higher intestinal bile acid content (see Chapter III) and a lack of appropriate nutrients.

A similar approach was taken by Jacobsen *et al.*,[54] who introduced a suicide plasmid into a wild-type *E. coli* to generally prevent the possibility of a genetically altered coli bacterium from spreading into the intestinal tract of humans and animals where *E. coli* would normally be present. They introduced a plasmid pPKL100 containing the suicide *hok* (host kill) gene into *E. coli* BJ4 originally isolated from a rat to produce *E. coli* BJ6(*hok*+). The latter organism established itself well in GF rats, but was quickly eliminated in the presence of other coli bacteria. Thus, introduction of the *hok* suicide gene would make the *in vivo* study of genetically altered *E. coli* and other suitable microorganisms possible while preventing their spread beyond the experimental study.

Materials With Antimicrobial Action

The above-cited examples of interaction of microorganisms, and especially the phenomenon of repression of certain bacteria by certain others, has given rise to speculation about growth retarding or possibly growth inhibiting factors. Early speculation had been about the production of volatile fatty acids,[38,39] but later research seemed not to confirm this.[45,46] Another possibility to be considered was the production of a specific compound with antibiotic-like properties that could have been produced, under favorable conditions, by one or a combination of intestinal microorganisms. Ducluzeau *et al.*[55,56] reported the production by *Bacillus licheniformis* of a bacitracin-like material which inhibited the growth of various *C. perfringens* sp. and in addition had an inhibitory effect on some other microorganisms. They also reported that certain oxygen-sensitive clostridia obtained from CV mice, in association with *E. coli* K-12, would create a barrier effect against *Shigella flexneri* comparable to that of the intact CV flora. They again mentioned the possibility of an antibiotic-like substance being formed under their experimental conditions.[57] This author is not aware of any active, well-defined substance having been isolated in this or possibly later studies. To what extent these putative antibiotic substances are involved in the earlier-mentioned phenomenon of CR is not clear at the moment.

Translocation Studies

As indicated earlier, the translocation of potential pathogens from the intestine is a major danger in all conditions that impair the integrity of the intestinal wall, or otherwise disrupt the intestinal ecosystem. Flanagan *et al.*[58] reported a much higher mortality after severe hemorrhagic shock in CV than in GF Sprague-Dawley rats. They were able to demonstrate the presence of translocated *E. coli* in the blood of the CV rats only 2 h after the onset of shock. A series of studies then investigated the conditions which promote translocation of microorganisms, notably potential pathogens, from the intestine. Translocation is presumed to occur via the lymph circulation and is generally measured by the number of viable organisms found in the mesenteric nodes. In the introduction to their 1979 publication, Berg and Garlington[59] briefly review the early work in this field. Their own data had shown that in SPF mice (SPF being a somewhat uncertain definition) no translation to the mesenteric nodes took place. However, when the total cecal flora of these animals was cultured and reintroduced via gavage into GF mice, sufficient disruption of the microbial ecosystem had taken place to allow translocation of the indigenous *E. coli* and lactobacilli. When the indigenous *E. coli* or *L. acidophilus* were introduced as monoassociated, translocation was even more pronounced, presumably because much larger numbers of organisms were established in the gut of the monoassociates. However, when GF mice were monassociated with an indigenous *E. coli* C25, which then translocated readily, this translocation could be totally inhibited by the concomitant introduction of the whole cecal flora of SPF mice, which drastically reduced the level of *E. coli* in the cecum.[60] A later study indicated that especially anaerobic bacteria play a pivotal role in limiting translocation.[61] Steffen and Berg[62] had already determined that under conditions of limited association when many bacterial strain readily translocate to the mesenteric nodes, they do so in proportion to the numbers that had been established in the cecum.

Berg,[63] whose group has done major work in this field, mentions disruption of the microbial ecosystem of the gut, e.g., by antibiotics, and a diminished immune potential as the main factors responsible for translocation. Once again, the gnotobiote would seem to be an ideal, totally controllable model to study the conditions under which translocation might take place or might be enhanced. On the other hand, the abnormal conditions created by the presence of only one or at best a limited number of selected microbial species, and the effect of this condition on the physiology of the gut wall (see Chapter III) may limit generalization of the results obtained under those conditions.

Nevertheless, some of the conditions which generally will promote translocation have become obvious: (1) a disturbance of the microbial ecosystem, e.g., by antibiotics, which will cause the overgrowth of a certain bacterial species in the gut; (2) a dysfunction of the immune system, as in the athymic mouse; (3) a disruption of the integrity of the intestinal wall, experimentally achieved, e.g., by oral administration of ricinoleic acid[64] or by irradiation.

Early studies in SPF mice had suggested that strictly anaerobic bacteria did not translocate unless T cell function was abrogated.[65] However, later work by Debure *et al.*[66] showed that in mono- and diassociated mice strictly anaerobic strains belonging to *Bacteroides, Clostridium,* and *Fusobacterium* did translocate to the mesenteric lymph nodes of immunologically intact mice. They saw the numbers that established in the cecum of these mono- and diassociated gnotobiotes as a main factor in determining the level of translocation. Still, it seems that when established as monoassociates in GN mice, Gram-negative bacilli translocated in large number, whereas Gram-positive bacteria did so at an intermediate level, and obligate anaerobes translocated only at very low levels.[67] A recent report by Wells et al.[68] claims, however, that S. faecalis translocates to the liver in greater numbers than either E. coli or P. mirabilis. In view of the fact that under stress situations especially the Gram-negative organisms tend to translocate, Youssef et al.[69] introduced human isolates of enteropathogenic Campylobacter sp. and of a number of E. coli strains as monoassociates in mice. No relation between their human pathogenicity and the level of translocation could be established.

A general stimulation of the immune system, achieved by i.p. injection of formalin-killed Proprionibacterium acnes sharply reduced the translocation of E. coli in GN mice.[70,71] The data imply that not only T cell function, but also the role of stimulated macrophages is important in restricting translocation. Earlier work by Berg and Garlington,[72] who vaccinated GF mice with a heat-killed strain of E. coli, had suggested that B cell protection via antibody formation did not substantially affect translation in animals subsequently monoassociated with E. coli.

In conclusion, it appears that the factors controlling translation are

1. Character of the organism;
2. Numbers, largely controlled by the sum total of the intestinal ecosystem;
3. Integrity of the intestinal wall;
4. Immune potential, with emphasis on T cell and macrophage function.

In the latter case it would appear that the macrophages are the effecter cells, and most likely are under T cell control.[73,74]

Berg[75] has covered all the above material in a recent review.

REFERENCES

1. du Vigneau, V., Ressler, C., Rachele, J. R., Reyniers, J. A., and Luckey, T. D., The synthesis of "biologically labile" methylgroups in the germfree rat, J. Nutr., 45, 361, 1951.
2. Orland, F. J., Blayney, J. R., Harrison, R. W., Reyniers, J. A., Trexler, P. C., Wagner, M., Gordon, H. A., and Luckey, T. D., J. Dent. Res., 33, 147, 1954.
3. Wagner, M., Determination of germfree status, Ann. N.Y. Acad. Sci., 78, 89, 1959.
4. Pollard, M. and Kajima, M., Leukemia in germfree rats, Proc. Soc. Exp. Biol. Med., 121, 585, 1966.
5. Dymsza, H. A., Stoewsand, G. S., Enright, J. J., Trexler, P. C., and Gall, L. C., Human indigenous microflora in gnotobiotic rats, Nature, 208, 1236, 1965.
6. Gibbons, R. J., Socransky, S. S., and Kapsimalis, B., Establishment of human indigenous bacteria in germfree mice, J. Bacteriol., 88, 1316, 1964.
7. Schaedler, R. W., Dubos, R., and Costello, R., Association of germfree mice with bacteria isolated from normal mice, J. Exp. Med., 122, 77, 1965.

8. Margard, W. L. and Peters, A. C., A study of gnotobiotic mice monocontaminated with *Salmonella typhimurium*, *Lab. Anim. Care*, 14, 200, 1964.
9. Knight, P. L., Influence of *Salmonella typhimurium* on ileum and spleen morphology of germfree rats, *Indiana Acad. Sci.*, 72, 78, 1964.
10. Wostmann, B. S., Pleasants, J. R., and Bealmear, P., Diet and immune mechanisms, in *Germfree Biology; Experimental and Clinical Aspects*, Mirand, E. A. and Back, N., Eds., Plenum Press, New York, 1969, 287.
11. Sasaki, S., Onishi, N., Nishakawa, T., Suzuki, R., Maeda, R., Takahashi, T., Usuda, M., Nomura, T., and Saito, M., Monoassociation with bacteria in the intestines of germfree mice, *Keio J. Med.*, 19, 87, 1970.
12. Sack, R. B. and Miller, C. A., Progressive changes of vibrio serotypes in germfree mice infected with *Vibrio cholerae*, *J. Bacteriol.*, 99, 688, 1969.
13. Shinamura, T., Immune response in germfree mice orally immunized with *Vibrio cholerae*, *Keio J. Med.*, 21, 113, 1972.
14. Shinamura, T., Tazumi, S., Hashimoto, K., and Sasaki, S., Experimental cholera in germfree suckling mice, *Infect. Immun.*, 34, 296, 1981.
15. Fox, J. G., Blanco, M., Murphy, J. C., Taylor, N. S., Lee, A., Kabok, Z., and Pappo, J., Local and systemic immune responses in murine *Helicobacter felis* active chronic gastritis, *Infect. Immun.*, 61, 2309, 1993.
16. Raibaud, P., Dickinson, A. B., Sacquet, E., Charlier, H., and Mocquot, G., Implantation controlée chez le rat gnotobiotique de différents genres microbiens isolés du rat conventionnel, *Ann. Inst. Pasteur*, 111, 193, 1966.
17. Ducluzeau, R., Équilibre entres deux souches bactériennes, *Escherichia coli* and *Staphylococcus pyogenes*, selon les conditions de leur ensemencement dans le tube digestif des souris axénique, *C. R. Acad. Sci. Paris*, 265, 1657, 1967.
18. Ducluzeau, R. and Raibaud, P., Influence de la vaccination sur l'implantation d'*Escherichia coli* et de *Staphylococcus pyogenes* dans le tractus digestif de la souris axénique, *Ann. Inst. Pasteur*, 114, 846, 1968.
19. Ducluzeau, R. and Galinha, A., Recombinaison *in vivo* entre une souche Hfr et une souche F- d'*Escherichia coli* L_{12} ensemencées dans le tube digestif de souris axénique, *C. R. Acad. Sci. Paris*, 264, 177, 1967.
20. Ducluzeau, R., Salomon, J. C., and Huppert, J., Ensemencement d'une souche lysogène et d'une souche sensible d'*Escherichia coli* K_{12} dans le tube digestif de souris axéniques: établissement d'un équilibre, *Ann. Inst. Pasteur*, 112, 153, 1967.
21. Ducluzeau, R., Le Garrec, D., and Salomon, J. C., Interaction entre une souche sensible et deux souches lysogènes de Staphylocoques ensemencées dans le tube digestif de souris axéniques, *Ann. Inst. Pasteur*, 114, 857, 1968.
22. Morotomi, M., Watanabe, T., Suegara, N., Kawai, Y., and Mutai, M., Distribution of indigenous bacteria in the digestive tract of conventional and gnotobiotic rats, *Infect. Immun.*, 11, 962, 1975.
23. Ducluzeau, R., Dubos, F., and Raibaud, P., Effet antagoniste d'une souche de *Lactobacillus* sur une souche de *Ristella* sp. dans le tube digestive de souris "gnotoxéniques" absorbant du lactose, *Ann. Inst. Pasteur*, 121, 777, 1971.
24. Nielsen, E. M., Schlundt, J., Gunvig, A., and Jacobsen, B. L., Epithelial, mucus and lumen subpopulations of *Escherichia coli* in the large intestine of conventional and gnotobiotic rats, *Microb. Ecol. Health Dis.*, 7, 263, 1994.
25. Ducluzeau, R. and Raibaud, P., Intérêt des systèmes gnotoxéniques pour l'étude des relations hôte-flore microbienne du tube digestif, *Reprod. Nutr. Dévelop.*, 20, 1667, 1980.
26. Raibaud, P., Ducluzeau, R., Dubos, F., Hudault, S., Bewa, H., and Muller, M. C., Implantation of bacteria from the digestive tract of man and various animals into gnotobiotic mice, *Am. J. Clin. Nutr.*, 33, 2440, 1980.
27. Gismondo, M. R., Drago, L., Lombardi, A., and Fassina, C., An experimental model to reproduce some bacterial intestinal cocultures in germfree mice, *Drugs Exp. Clin. Res.*, 20, 149, 1994.

28. Corthier, G., Dubos, F., and Raibaud, P., Modulation of cytotoxin production by *Clostridium difficile* in the intestinal tract of gnotobiotic mice inoculated with various human intestinal bacteria, *Appl. Environ. Microbiol.*, 49, 250, 1985.

29. Miller, C. E. and Feeley, J. C., Competitive effects of intestinal microflora on *Vibrio cholerae* in gnotobiotic mice, *Lab. Anim. Sci.*, 25, 454, 1975.

30. Schlund, J., Saadbye, P., Lohmann, B., Jacobsen, B. L., and Nielsen, A. M., Conjugal transfer of plasmid DNA between *Lactobacillus lactis* strains and distribution of transconjugants in the digestive tract of gnotobiotic rats, *Microb. Ecol. Health Dis.*, 7, 59, 1994.

31. Duval-Iflah, Y., Gainche, I., Oriet, M., Lett, M., and Hubert, J., Recombinant DNA transfer to *Escherichia coli* of human fecal origin *in vitro* and in the digestive tract of gnotobiotic mice, *FEMS Microbiol. Ecol.*, 15, 79, 1994.

32. Phillips, A. and Balish, E., Growth and invasiveness of *Candida albicans* in the germfree and conventional mouse, *Appl. Microbiol.*, 14, 737, 1966.

33. Nishikawa, T., Hatano, H., Onishi, N., Sasaki, S., and Nomura, T., The establishment of *Candida albicans* in the elementary tract of the germfree mice, and antagonism with *Escherichia coli* after oral inoculation, *Jpn. J. Microbiol.*, 13, 263, 1969.

34. Rogers, T. and Balish, E., Experimental *Candida albicans* infection in conventional mice and germfree rats, *Infect. Immun.*, 14, 33, 1976.

35. Balish, E., Filutowicz, H., and Oberley, T. D., Correlates of cell-mediated immunity in *Candida albicans*-colonized gnotobiotic mice, *Infect. Immun.*, 58, 107, 1990.

36. Balish, E., Jensen, J., Warner, T., Brekke, J., and Leonard, B., Mucosal and disseminated candidiasis in gnotobiotic SCID mice, *J. Med. Vet. Mycol.*, 31, 143, 1993.

37. Raibaud, P., Ducluzeau, R., Dubos, F., and Sacquet, E., Spore formation and germination of *Clostridium perfringens* in the digestive tract of holoxenic and axenic mice, *J. Appl. Bacteriol.*, 35, 177, 1972.

38. Van der Waaij, D., Berghuis de Vries, J. M., and Lekkerkerk-van der Wees, J. E. C., Colonization resistance of the digestive tract in conventional and antibiotic-treated mice, *J. Hyg. (Cambridge)*, 69, 405, 1971.

39. Van der Waaij, D., The colonization resistance of the digestive tract in man and animals, in *Clinical and Experimental Gnotobiotics. Proceedings of the VI International Symposium on Gnotobiology*, Fliedner, T. L. *et al.*, Eds., Gustav Fischer, New York, 1979, 155.

40. Koopman, J. P., Welling, G. J., Huybregts, A. W. M., Mullink, J. W. M. A., and Prins, R. A., Association of germfree mice with intestinal microfloras, *Z. Versuchstierk.*, 23, 145, 1981.

41. Raibaud, P., Ducluzeau, R., Dubos, F., Hudault, S., Bewa, H., and Muller, M. C., Implantation of bacteria from the digestive tract of man and various animals into gnotobiotic mice, *Am. J. Clin. Nutr.*, 33, 2440, 1980.

42. Lusk, R. H., Fekety, R., Silva, J., Browne, R. A., Ringgler, D. H., and Abrams, G. D., Clindamycin-induced enterocolitis in hamsters, *J. Infect. Dis.*, 137, 464, 1978.

43. Su, W. J., Bourlioux, P., Bournaud, M., Besnier, M. O., and Fourniat, J., Transfer de la flore caecale du hamster à la souris C3H axénique: utilisation de cette modèle pour étudier la flore du barrière anti-*Clostridium difficile*, *Can. J. Microbiol.*, 32, 132, 1986.

44. Wilson, K. H. and Freter, R., Interaction of *Clostridium difficile* and *Escherichia coli* with microfloras in continuous flow-cultures and gnotobiotic mice, *Infect. Immun.*, 54, 354, 1986.

45. Guiot, H. F. L., Volatile fatty acids and the selective growth inhibition of aerobic bacteria in the guts of rats, in *Recent Advances in Germfree Research. Proceedings of the VII International Symposium on Gnotobiology*, Sasaki, S. *et al.*, Tokai University Press, Tokyo, 1981, 219.

46. Su, W. J., Waechter, M. J., Bourlioux, P., Dolegeal, M., Fourniat, J., and Mahuzier, G., Role of volatile fatty acids in colonization resistance to *Clostridium difficile* in gnotobiotic mice, *Infect. Immun.*, 55, 1686, 1987.

47. Yurdusev, N., Nicolas, J. L., Ladiré, M., Ducluzeau, R., and Raibaud, P., Antagonistic effect exerted by three strictly anaerobic strains against various strains of *Clostridium perfringens* in gnotobiotic rodent intestines, *Can. J. Microbiol.*, 33, 226, 1987.

48. Boureau, H., Decré, D., Popoff, M., Bertocci, A., Su, W. J., and Bourlioux, P., Isolation and identification of microflora resistant to colonization by *Clostridium difficile*, *Microecol. Ther.*, 18, 117, 1988.

49. Boureau, H., Salanon, C., Decaens, C., and Bourlioux, P., Caecal localization of the specific microbiota resistant to *Clostridium difficile* colonization in gnotobiotic mice, *Microb. Ecol. Health Dis.*, 7, 111, 1994.

50. Yurdusev, N., Ladire, M., Ducluzeau, R., and Raibaud, P., Antagonism exerted by an association of a *Bacteroides thetaiotaomicron* strain and a *Fusobacterium necrogenes* strain against *Clostridium perfringens* in gnotobiotic mice and in fecal suspensions incubated *in vitro*, *Infect. Immun.*, 57, 724, 1989.

51. Freter, R., Abrams, G. D., and Aranki, A., Patterns of interaction in gnotobiotic mice among bacteria of a synthetic "normal" intestinal flora, in *Germfree Research: Biological Effects of Gnotobiotic Environments. Proceedings of the IV International Symposium on Germfree Research*, Heneghan, J. B., Ed., Academic Press, New York, 1973, 429.

52. Wells, C. L., Johnson, W. J., Kan, C. M., and Balish, E., Inability of debilitated *Escherichia coli* chi 1776 to colonize germfree rodents, *Nature*, 274, 397, 1978.

53. NIH, Guidelines for Research Involving DNA Molecules, U.S. Department of Health, Education and Welfare, Public Health Service, National Institutes of Health, Bethesda, Maryland, October 1977.

54. Jacobsen, B. L., Schlundt, J., and Fischer, G., Study of a conditional suicide system for biological containment of bacteria in germfree rats, *Microb. Ecol. Health Dis.*, 6, 109, 1993.

55. Ducluzeau, R., Dubos, F., Raibaud, P., and Abrams, G. D., Inhibition of *Clostridium perfringens* by an antibiotic substance produced by *Bacillus licheniformis* in the digestive tract of gnotobiotic mice: effect on other bacteria from the digestive tract, *Antimicrob. Agents Chemother.*, 9, 20, 1976.

56. Ducluzeau, R., Dubos, F., Raibaud, P., and Abrams, G. D., Production of an antibiotic substance by *Bacillus licheniformis* within the digestive tract of gnotobiotic mice, *Antimicrob. Agents Chemother.*, 13, 97, 1978.

57. Ducluzeau, R., Ladire, M., Callut, C., Raibaud, P., and Abrams, G. D., Antagonistic effect of extremely oxygen-sensitive Clostridia from the microflora of conventional mice and of *Escherichia coli* against *Shigella flexneri* in the digestive tract of gnotobiotic mice, *Infect. Immun.*, 17, 415, 1977.

58. Flanagan, J. J., Rush, B. F., Murphy, T. F., Smith, S., Machiedo, G. N., Hsieh, J., Rosa, D. M., and Heneghan, J. B., A "treated" model for severe hemorrhagic shock: a comparison of conventional and germfree animals, *J. Med.*, 21, 104, 1990.

59. Berg, R. D. and Garlington, A. W., Translocation of certain indigenous bacteria from the gastrointestinal tract to the mesenteric lymph nodes and other organs in a gnotobiotic mouse model, *Infect. Immun.*, 23, 403, 1979.

60. Berg, R. D. and Owens, W. E., Inhibition of translocation of viable *Escherichia coli* from the intestinal tract by bacterial antagonism, *Infect. Immun.*, 25, 820, 1979.

61. Wells, C. L., Maddaus, M. A., Jechorek, R. P., and Simmons, R. L., The role of intestinal anaerobic bacteria in colonization resistance, *Eur. J. Clin. Microbiol. Infect. Dis.*, 7, 107, 1988.

62. Steffen, E. K. and Berg, R. D., Relationship between cecal population levels of indigenous bacteria and translocation to the mesenteric lymph nodes, *Infect. Immun.*, 39, 1252, 1983.

63. Berg, R. D., Factors influencing the translocation of bacteria from the gastrointestinal tract, in *Recent Advances in Germfree Research. Proceedings of the VII International Symposium on Gnotobiology*, Sasaki, S. *et al.*, Eds., Tokai University Press, Tokyo, 1981, 411.

64. Berg, R. D., Bacterial translocation from the intestine, *Jikken Dobutsu (Exp. Animals)*, 34, 1, 1985.

65. Owens, W. E. and Berg, R. D., Bacterial translocation from the gastrointestinal tract of athymic (nu/nu) mice, *Infect. Immun.*, 27, 461, 1980.

66. Debure, A., Rambaud, J.C., Ducluzeau, R., Yurdusev, N., and Raibaud, P., Translocation of strictly anaerobic bacteria from the intestinal tract to the mesenteric lymph nodes in gnotobiotic rodents, *Ann. Inst. Pasteur*, 138, 213, 1987.

67. Steffen, E. K., Berg, R. D., and Deitch, E. A., Comparison of translocation rates of various indigenous bacteria from the gastrointestinal tract to the mesenteric lymph node, *J. Infect. Dis.*, 157, 1032, 1988.

68. Wells, C. L., Jechorek, R. P., and Gillingham, K. J., Relative contributions of host and microbial factors in bacterial translocation, *Arch. Surg.*, 126, 247, 1991.

69. Youssef, M., Corthier, G., Goossens, H., Tancrede, C., Henry Amar, M., and Andremont, M., Comparative translocation of enteropathogenic *Campylobacter* spp. and *Escherichia coli* from the intestinal tract of gnotobiotic mice, *Infect. Immun.*, 55, 1019, 1987.

70. Fuller, K. G. and Berg, R. D., Inhibition of bacterial translocation from the gastrointestinal tract by nonspecific immune stimulation, in *Germfree Research: Microflora Control and its Application to the Biomedical Science. Proceedings of the VIII International Symposium on Germfree Research*, Wostmann, B.S., Ed., Alan R. Liss, New York, 1985, 195.

71. Gautreaux, M. D., Deitch, E. A., and Berg, R. D., Immunological mechanisms preventing bacterial translocation from the gastrointestinal tract, *Microecol. Ther.*, 20, 31, 1990.

72. Berg, R. D. and Garlington, A. W., Translocation of *Escherichia coli* from the gastrointestinal tract to the mesenteric lymphnodes in gnotobiotic mice receiving *Escherichia coli* vaccines before colonization, *Infect. Immun.*, 30, 894, 1980.

73. Wells, C. L., Jechorek, R. P., Feltis, B. A., and Erlandsen, S. L., *Escherichia coli, Proteus mirabilis*, and *Enterococcus faecalis*: uptake and survival within mouse peritoneal macrophages, uptake by cultured epithelial cells, and bacterial translocation in monoassociated mice, in Abstr. 30th Annu. Meet. Assoc. Gnotobiotics, Madison, Wisconsin, 1992, Abstr. # 26.

74. Gautreaux, M. and Berg, R. D., T cells and macrophages in the host defense against bacterial translocation, in Abstr. 30th Annu. Meet. Assoc. Gnotobiotics, Madison, Wisconsin, 1992, Abstr. # 27.

75. Berg, R. D., Bacterial translocation from the gastrointestinal tract, *J. Med.*, 23, 217, 1992.

Chapter XI

APPLICATIONS: PAST, PRESENT, AND FUTURE. PART II. USE OF THE GNOTOBIOTE IN THE STUDY OF DISEASE

THE AGING SYNDROME

"Endpoint Studies"

The GF animal makes it possible to conduct clean "endpoint studies": no uncontrolled microbial infection will affect the disrupted or declining functional capabilities of the animal under study. It was thus recognized early on that the GF animal, and for that matter also the GN animal, are models of choice for studies in, e.g., radiation, lethal trauma, and aging.

The first descriptions of the causes of death in the absence of a bacterial microflora were given by Gordon *et al.*[1] and by Walburg and Cosgrove,[2] both groups following GF mice until natural death. These studies, detailed in Chapter II, would at present be considered potentially flawed because although these mice were free of bacteria, they definitely were not free of viruses. The more recent Lobund Aging Study, using GF Lobund-Wistar (L-W) rats (see Chapters II and IX), does meet this more stringent requirement of germfreeness. Whereas this study confirmed the longer life span of *ad libitum*-fed germfree mice, it also showed that as far as longevity is concerned, limiting dietary intake is an overriding factor in the aging syndrome. Restricted to 12 g of diet per day, survival curves of both GF and CV rats became almost identical, with the 50% survival age for both groups being 37.2 months (Chapter II, Figure 3).

Pathology at natural death was not always clearly defined, especially since a casein-free natural ingredient diet like L-485[3] was found to greatly reduce the earlier observed kidney pathology in older animals, both GF and CV. Cancer proved to be a major factor, as indicated in Chapter IX, with many of the tumors endocrine related. In general, tumor development appears to be slower in GF rats and mice. Deerberg and Kaspareit[4] reported a lower occurrence of endometrial carcinomas in GF than in CV female BDII/Han rats. This phenomenon could be related to a shift in hormonal homeostasis due to the absence of microbial effects on the steroid hormones circulating in the enterohepatic circulation (see Chapter II, papers by Gustafsson *et al.*).

Lobund Aging Study

Since the early studies by McCay *et al.*[5] the scientific world has been intrigued by the effects of dietary restriction on health and longevity. After World War II, Van der Rijst *et al.*[6] showed that in rats a 30% reduction in *ad libitum* dietary intake resulted in a lean, healthy animal with a substantially increased life span. More recent studies by Masoro *et al.*[7] and many other workers have confirmed this phenomenon. However, in the early studies on aging, evaluation of the final phase was always hampered by the possibility of bacterial infection limiting the life span and thereby obscuring the true course of the physiological aging process. Hence the Lobund Laboratory started the Lobund Aging Study in the early 1980s using the GF L-W rat as its experimental model.[8]

Specific topics of the Lobund Aging Study have been incorporated in Chapters I, II, IV, V, VII, and IX. Much of the material of this study has been brought together in a monograph entitled "Dietary Restriction and Aging".[9]

CANCER RESEARCH

The occurrence of cancer as a potentially life-threatening malignancy in older GF and CV rats and mice was discussed in Chapter IX. Once it had been established that cancer does occur in GF animals, although not always to the same extent, this suggested the animal to be an obvious model for the controlled study of the factors which affect its etiology and pathogenesis. Here the emphasis will be on the differences between the GF and the CV state as far as they affect the action of materials, either naturally occurring or purposely administered, which will induce or enhance cancer.

As early as 1964, Laquer, and later Spatz *et al.*,[10,11] had reported that cycasin, a cancer-causing agent found in the cycad nut, did not cause cancer when fed to GF rats nor did it produce the characteristic liver tumors when given by i.p. injection to CV rats. As it turned out, the actual carcinogen was methylazoxymethanol (MAM), which occurs as a glycoside in cycasin and is liberated by bacterial action in the gut.[12] Later, Iwasaki et al.[13] studied the cycasin aglucon acetate (MAM acetate) in GF mice and found its carcinogenic action in the large intestine to be enhanced not only by the addition of lard to the diet, but also after monoassociation with *Lactobacillus arabinosus*. The enhancement by the addition of lard, bringing the total fat content of the diet to almost 18%, may be explained by a putative higher bile acid concentration in the lower gut and its action on the local mucosa. However, the action of the lactobacillus may indicate that additional MAM metabolites may have been formed which contributed to its carcinogenic potential.

The inducer(s) of colon cancer have been a matter of debate for many years, and many workers blame very small amounts of man-made substances that have entered the food chain. Both Weisburger et al.[14] and Balish et al.[15] have described the induction of colon cancer by intrarectal injection of N-methyl-N'-nitro-N-nitrosoguanidine (MNNG) in the GF rat. Both groups concluded that the GF rat was possibly more susceptible than its CV counterpart. Earlier studies, however, had suggested that microbially modified intestinal steroids may be potentially involved in the causation of colon cancer. Thus, in 1978 Reddy et al.[16] used the GF rat to evaluate the promoting effect of deoxycholic acid, one of the major bacterially modified intestinal bile acids in humans, on the occurrence of colon adenocarcinomas after intrarectal instillation of MNNG. They reported that deoxycholic acid alone administered intrarectally in relatively high doses (3 g per rat per week for 1 year) was without effect, but in combination with early administration of MNNG this amount caused a significant increase in the number of tumors found 1 year later. Deoxycholic acid thus indeed had a promoting effect, but only in concentrations higher than physiologically possible.

Just as the GF model was needed to conduct a controlled study of the effect of deoxycholic acid, an intestinally produced bacterial product of cholic acid, other workers used the GF animal to evaluate the effects of the microflora on the action of carcinogens and their systemic metabolites. Weisburger et al.[17] were among the first to report that because of enzymatic and/or reductive action of the intestinal microflora, potential carcinogens may be changed in the enterohepatic circulation. They reported extensive glucuronidization and sulfatation of the carcinogen N-hydroxy-N-2-fluorenylacetamide (N-OH-FAA) in GF, but little in CV rats where these hepatic detoxification products had been exposed to microbial action.

Various carcinogens carry nitro groups, and their action appears to be enhanced by microbial reduction to the corresponding amino group. El Bayoumy et al.[18,19] analyzed feces and urine of GF and CV rats after gavage of 1-nitropyrene, an aromatic four-ring system, and detected 1-aminopyrene only in the CV animals, indicating that in this case the microflora was indeed involved in the reduction of the nitro group. Morton and Wang,[20] on the other hand, studied the action of the bladder carcinogen N-[4-(5-nitro-2-furyl)-2-thiazolyl]-formamide (FANFT), a compound much different from the four-ring aromates, and reported that after p.o. administration of the ^{35}S-labeled material the liver, kidney, and bladder of the GF rat retained more of the label than its CV counterpart. Binding to protein was relatively greater than binding to RNA, while DNA appeared to bind little of the labeled material. Here, the data seem to suggest that part of FANFT was microbially altered in the CV gut to the extent that its carcinogenic binding potential may have been decreased.

In 1990 Delclos et al.[21] reported the reduction of another nitro compound of a four-ring aromatic system, (^{3}H-labeled) 6-nitrochrysene, and found that the reduction to 6-aminochrysene took place in GF as well as in CV mice, although in GF mice to only 25% of the CV amount. Their data indicate that in this case the liver played a role in the conversion of 6-nitrochrysene to, presumably, a number of metabolites. In the final analysis, the levels of carcinogen-DNA adducts in the lungs and livers of GF and CV mice proved to be similar.

All this pointed to the necessity of establishing which of these conversions were basically systemic and which were brought about by the intestinal microflora. In a recent publication Yang et al.[22] give an extensive description of the many neutral metabolites of benzo[a]pyrene in the urine of the GF rat. The data indicate the extent to which the liver and other body tissues will go to eliminate offensive materials, achieving this to a great extent by conjugation. This was also demonstrated by the work of George et al.,[23] who studied the mutagenicity of urine metabolites of GF and conventionalized Fischer rats after per os administration of the promutagen 2,6-dinitrotoluene. They concluded that although the intestinal microflora plays an important role in the conversion of 2,6-dinitrotoluene to mutagenic material, small-intestinal mucosal and/or hepatic enzymes obviously contribute to the generation of genotoxicants.

Pollard[24] was the first to describe the spontaneous occurrence of prostate adenocarcinoma in a specific strain of rat. Thus far the L-W rat is the only strain in which this tumor develops in the older male, making it a model of choice for the study of a disease prevalent in the older human male. GF rats maintained on natural ingredient diet L-485 showed a 10% occurrence of prostate adenocarcinomas at the average age of 34 months, against 26% in comparable CV rats before the age of 30 months.[25,26] Again we see here that these endocrine-related tumors develop slower in the GF than in the CV animal. This could be related to the fact that serum testosterone concentrations seem to be always slightly lower in the ad libitum-fed GF L-W rat than in its CV counterpart (Chapter IV, Table 6). In diet-restricted GF rats, however, testosterone levels tend to be substantially higher than in the ad libitum-fed GF animals (Figure 1) whereas the occurrence of prostate tumors within a specific age group is definitely less (see Chapter IX).

In Chapter VII the possibility of xenogeneic bone marrow transplants into GF and GN rats was discussed. In a similar vein, Zimmerman et al.[27]

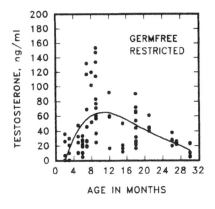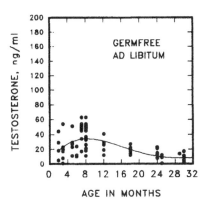

FIGURE 1
Serum testosterone concentrations of *ad libitum*-fed and diet-restricted germfree Lobund-Wistar rats.

recently reported that they had succeeded in inoculating material from human colorectal carcinomas into young athymic gnotobiotic mice. Two of the six tumor preparations established well in the lungs of the recipient GN mice, thereby providing another model for the study of this disease. The authors stressed the potential importance of low NK cell activity in these mice.

ROLE OF THE MICROFLORA IN STEROID METABOLISM AND CARDIOVASCULAR DISEASE

The potential role of cholesterol in cardiovascular disease (CVD) has been suspected since cholesterol-containing fatty streaks were found in the arteries of young American battlefield casualties during the Korean war. Soon thereafter, studies were started on the influence of intestinal bacteria on cholesterol and bile acid metabolism. Most of this research is reviewed in Chapters IV and V. Here, it should be stressed again that for the purpose of this research neither the rat nor the mouse, which have fairly comparable steroid metabolisms, are models of choice. Of the gno-tobiotic models thus far available, only the Mongolian gerbil can be con-sidered to have a cholesterol and bile acid metabolism that is reasonably comparable to that of the human. Unlike the rat and mouse, the gerbil does not produce β-muricholic acid as a primary bile acid. In the rat and mouse the microflora will convert β-muricholic acid via hyodeoxycholic acid to ω-muricholic acid. These bile acids are much more easily removed from the body, thus giving these rodents a natural protection against cholesterol overload (see Chapter IV). The gerbil does not have this po-tential and, consequently, is much more sensitive to excess cholesterol intake[28,29] and must be considered a model of choice for these studies (for more details see Chapter IV).

Another matter that needs investigation is the influence the intestinal microflora may have on the endocrine system, since some of its members will enter the enterohepatic circulation and thus be affected by the microflora. An example of this can be found in a series of publications by Gustafsson and co-workers at the Karolinska Institute under the general title "Steroids in Germfree and Conventional Rats" published in the *European Journal of Biochemistry* between 1966 and 1970 (see Chapter II). Although the implications of these studies were, to the best of this author's knowledge, never followed up, this obviously will have to be an important field of study in the future.

DENTAL CARIES AND PERIODONTAL DISEASE STUDIES

Dental Caries

Ever since it became obvious that in the GF state animals do not develop dental caries,[30] the GF rat has been an obvious tool to study the role of specific microorganisms, and combinations of microorganisms, in the etiology of dental caries. Shortly after their first observations, Orland *et al.*[31] reported the production of dental caries in GN rats associated with an enterococcus and, in one case also a proteolytic bacillus, and in another an "anaerobic pleomorphic bacterium." Somewhat later, Fitzgerald *et al.*[32] described the production of dental caries in GN rats harboring a single strain of oral streptococcus isolated from CV rats. This established streptococci, rather than the often-assumed lactobacilli, as the main culprits in the etiology of dental caries[33] although some reports do indicate cariogenic effects of certain lactobacilli.[34,35] Kolenbrander *et al.*[36] recently again pointed to the streptococci as the "first colonizers" among the plethora of oral microorganisms during plaque formation in humans.

In 1972 Mikx *et al.*[37] mentioned that the list of cariogenic microorganisms now include streptococci, lactobacilli, and actinomyces sp., with *Streptococcus mutans* appearing as the most cariogenic. Although under CV rat and human conditions the oral flora consists of many species, they considered it essential to study at least simple combinations of organisms. They diassociated GF rats with both a cariogenic streptococcus (*S. mutans* or *S. sanguis*) and with *Veillonella alcalescens*, another organism inhabiting the mouth. Using a 16% sucrose diet, the usual favorite for this research, they observed a distinct dilution of the cariogenic potential of these streptococci by the second associate. The same group then conventionalized rats originally monoassociated with *S. mutans*, and again found a substantial dilution of the cariogenic effect of *S. mutans*. They imply, but do not show, that the resulting caries in the diassociated rats was still more severe than in comparable CV counterparts.[38]

The rat studies indicate that streptococci, notably S. mutans, but to an extent also lactobacilli, are major originators of dental caries in that animal. Van Houte and Russo[39] monoassociated Sprague Dawley rats with strains of S. mitis, S. sanguis, and S. mutans and checked colonization on the molar teeth. They report an adherence of viable cells in the order of 10^6, 10^7, and 10^8, respectively, again pointing to S. mutans as potentially the most cariogenic organism. The study also confirmed that sucrose is more cariogenic than glucose, although the latter statement has drawn criticism from Rosen,[40] whose data suggest that glucose might be more cariogenic in monoassociated rats.

More recently, Wilcox et al. have done extensive studies on the relationship between plaque-forming ability and cariogenicity of S. oralis[41] and of four species of S. mutans.[42] They state that plaque-forming ability correlated only loosely with cariogenicity. Harris et al.,[43] on the other hand, have suggested a closer relationship between plaque build-up and caries. They assume that glycogen-like polysaccharide formation on the tooth surface is the initiating factor in plaque formation and showed that a glycogen synthesis-deficient mutant of S. mutans had a significantly reduced cariogenic potential. However, Barletta et al.[44] had earlier studied a mutant of S. mutans serotype c gtfa which did not contain the gene for the production of small glucan molecules. They found this variant fully virulent in both GF and CV rats, and concluded that other glucan-forming enzyme systems might be the essential ones in the initiation of dental caries. Fitzgerald et al.[45] then studied the mechanisms of cariogenicity of S. mutans serotype c with a lactate dehydrogenase-deficient mutant of a human isolate and found it to be much less cariogenic in Sprague-Dawley rats than the wild strain. They speculated about the possibility of inoculating such a mutant strain with far less lactic acid-producing capacity to counteract the cariogenicity of the wild-type S. mutans.

The fact that dental caries obviously is a bacterial disease logically drew attention to the possibility of immunological protection by introducing the culpable organism in such a way that high saliva antibody titers might be obtained. Wagner,[46] using one S. mutans, two S. sanguis-like, and one L. casei strain injected the homologous formalin-killed organisms subcutaneously in L-W rats that had been orally monoassociated with these bacteria. In all cases but one saliva titers were approximately ten times higher in the treated than in the nontreated gnotobiotes. Caries was significantly reduced, and the caries incidence proved to be inversely related to the specific agglutination antibody titer attained in the saliva. McGhee et al.[47] essentially followed the same procedure with a preparation of S. mutans, but injected it into the submandibular region, thus obtaining even higher saliva antibody titers. They found the antibody to be of the IgA type, which now comprised about half of the salivary immune globulin. The overall caries scores showed 70% reduction in the treated animals. In a later study this group immunized GF rats by local immunization with a ribosomal preparation from *S. mutans*. This afforded greater protection against caries formation than treatment with whole-cell preparations, and supposedly avoided the triggering of autoimmune reactions potentially resulting in heart and kidney pathology.[48] With this a real possibility for the treatment of dental caries seemed indicated, but for various reasons it was never translated into an effective treatment for humans.

Periodontal Disease

The first mention of periodontal disease in GF animals dates back to 1959, when Baer and Newton[49] described periodontal changes in GF Swiss-Webster mice maintained on diet L-356 (for diet L-356 see Chapter V) at the Lobund Laboratory. They observed these changes starting at 4 months of age, and reported that at the age of 6 months or more all of the observed animals showed periodontal disease. In a follow-up study with ex-breeders aged 12 to 30 months they found periodontal disease in all GF and CV animals, with no obvious difference in incidence or severity between the two. They concluded that living bacteria did not seem to be the primary etiologic agent in "this form of periodontal disease."[50] Then in 1981 Taubman *et al.*[51] observed the syndrome in the most defined model available, the GF-CD rat. They reported that GF rats maintained on the Lobund "antigen-free" chemically defined (CD) diet L-489E8 (for CD diets see Chapter VI) showed increased periodontal bone loss starting at 2 months after 1 mg of ovalbumin (OVA) had been added to the CD diet. Within 3 weeks after the start of OVA feeding, gingival lymphocyte numbers increased over those of the non-OVA controls. For the first 2 months T cells dominated among the gingival lymphocytes, but at 4 months T cells had decreased and B cells dominated by a factor of 6. Also, IgA appeared in intestinal perfusates after 9 days and in saliva after 23 days. The authors concluded that stimulation by OVA activated local immune phenomena caused the increased periodontal bone loss.

Notwithstanding the above, a number of gnotobiotic studies have been done with specific organisms thought to be actively involved in the etiology of periodontal disease. These include *Actinomyces, Bacteroides, Eikenella,* and *Porphyromonas* sp. In many cases monoassociation was combined with local application of the organism, which generally resulted in a more pronounced lesion. The question remains, however, to what extent the observed lesions resulted from action specific for the organism under study, or from the general activation of immune mechanisms by an organism with a preponderant action on the gingival region.

RADIATION BIOLOGY: ROLE OF THE MICROFLORA AND EFFECT OF BACTERIAL LIPOPOLYSACCHARIDES (LPS)

As mentioned in Chapter VII, a more extensive study of the effects of radiation was started because of the fear of a nuclear encounter or of nuclear accidents. However, later studies focused to a great extent on its use in organ transplantation. Early on, Reyniers et al.[52] had shown that the GF state imparted a certain amount of protection from X-radiation in rats. Later, McLaughlin et al.[53] reported that at X-ray levels of from 550 to even 40,000 rad GF Swiss-Webster mice always survived longer than their CV counterparts. Mice monoassociated with Escherichia coli survived longer than CV mice but not as long as the GF animals, an early suggestion of the possible involvement of LPS. Another such indication came from the studies by Onoue et al.[54] who, using 2 krad whole-body gamma irradiation of GF, GN, and CV ICR mice, found that monoassociation with E. coli or a Pseudomonas sp. brought survival times down to the CV range. Association with a Clostridium sp., surprisingly, increased survival time substantially. Although bacterial invasion was observed in the Pseudomonas and E. coli association and in the CV animals, they ascribed the cause of death to hematopoietic damage.

The early radiation studies have been summarized by van Bekkum.[55] They convincingly show the generally negative effects of the presence of a CV microflora. Walburg et al.[56] reported that for various strains of mice exposed to whole-body X-rays the LD_{50} dose was always higher for GF than for CV animals, but under comparable conditions this difference varied from 147 rad for RFM males to 34 rad for CF #1 mice (the CF strain originating from Carworth Farms, New York). Their data suggest that females might be more radiation resistant than males. They also observed that the LD_{50} of the various strains varied only half as much in GF as in CV mice, suggesting an effect of the differences in microflora and microflora products of the various strains.

Although the aforementioned studies appeared to implicate microbial products as a cause of the reduced radiation resistance of the CV animal when compared to its GF counterpart, in the absence of a microbial flora it was possible to show that a bacterial product like LPS was able to impart a certain protective effect. Since in GF mice this effect was maximal when LPS was injected about 24 h before irradiation, Smith et al. concluded that in this case LPS induced a marked but transient increase in bone marrow cells which could then compensate for the cells damaged by subsequent irradiation. This then would lead to a faster recovery of the animals.[57] A similar conclusion had been reached by Ledney and Wilson[58] who reported an increase of the LD_{50} X-ray dose of approximately 100 rad in both GF and CV CFW mice 24 h after receiving a 10-μg i.p. dose of LPS (GF 829 to 942, CV 739 to 834). The fact that this protection was limited to the "hematopoietic death range" appears to confirm its effect on the precursor bone marrow cells. Later, Wilson and Matzusawa[59] explained the protective effect of LPS "... by a mechanism that reduces the pO_2 of radiosensitive tissues to a level associated with protection" because of decreased blood flow in capillary beds resulting from vasoconstriction. Thus, LPS may ameliorate the effects of radiation, but its continuous influx, presumably caused by the damage to intestinal tissue, appears to overshadow its positive effect.

THE SHOCK SYNDROME: POTENTIAL INFLUENCE OF LPS

Translocation of Gram-negative organisms from the gut and subsequent release of LPS into the circulation appears to occur in all forms of shock.[60] Since the use of GN animals in these studies must be foreseen, it is important to first consider how the GF animal, obviously the animal at one end of the association spectrum, is able to handle LPS. During shock, as in the above-mentioned irradiation syndrome, increased permeability of the gut wall and translocation of microorganisms can be expected. Since all CV animals harbor Gram-negative organisms, their susceptibility to LPS could be affected by immunological mechanisms because of constant exposure. The GF animal would lack this exposure to LPS-producing organisms. In the case of GF rats and GF mice and of many of the gnotobiotes, we also have to consider their enlarged ceca with their output of materials that influence vascular and cardiac function (see Chapter IV). Heneghan[61] has concluded that at least for the study of the effects of the microflora in hemorrhagic shock, the cecectomized GF rat should be the model of choice.

Sensitivity to LPS was first studied by Landy et al.[62] in GF mice. Upon i.v. administration they found no significant differences with CV controls in the various reaction parameters they tested. Jensen et al.,[63] on the other hand, stated that GF mice were less susceptible to LPS from E. coli. They assumed that the LPS used by Landy et al., being derived from Salmonella enteritidis and from Shigella flexneri, organisms not indigenous to the mouse, might be the reason for the difference. However, Parant et al.[64] also using an LPS derived from S. enteritidis could find no difference in tolerance between GF and CV mice. It would appear that thus far no real differences have become obvious in the way GF and CV rodents handle LPS. To what extent this can be extrapolated to the GN animals, which take their place between the GF and CV models, will remain a matter of debate. Some potential effects on B cell, T cells, and macrophage function follow below. Recently, an LPS-binding protein was described that is secreted by the liver during an acute-phase response.[65]

Bulanda et al.[66] have studied the toxicity of staphylococcal toxic shock syndrome toxin 1 for GF and CV piglets. They found the GF piglets less susceptible than the corresponding CV controls, but ascribe the difference to a potentiating effect of LPS originating from the intestinal microflora of the CV animals.

Hemorrhagic Shock

The first hemorrhagic shock studies were carried out in 1957 by Zweifach et al.[67] with GF and CV rats. They saw no difference in the response to bleeding, in the duration of the hypotensive episode, or in the subsequent pathological changes. A few years later McNulty and Linares[68] reported essentially similar results. Both studies used animals obtained from the Lobund Laboratory at Notre Dame. Pentobarbital anesthesia was used. Then in 1967 Heneghan[69] described techniques that made it possible to use unanesthetized rats. He used GF and CV Sprague-Dawley rats, and again obtained comparable results. Later, Yale and Torhorst[70] used methoxyflurane as a general anesthetic to determine critical bleeding volume. This resulted in better survival, but again did not show a difference between GF and CV Sprague-Dawley rats.

In the meantime Heneghan and co-workers,[71,72] mindful of the aforementioned fact that materials originating in the enlarged cecum of the GF rat will influence hemodynamics, had started studying hemorrhagic shock in GF beagles — animals that do not show cecal enlargement in the GF state. Here they found indications that the GF beagle was more resistant to terminal hemorrhage than its CV counterpart. This led Heneghan's group to repeat the rat studies with cecectomized GF animals, which now showed the definite advantage of the absence of an intestinal microflora. Their data indicate that within 2 h the CV rats showed bacterial invasion of the bloodstream, but also that other factors besides bacteria and their products play a role in hemorrhagic shock.[73]

Thermal Shock

The difficulty of controlling infection and shock in severely burned patients led to these studies, which were all carried out under deep anesthesia by scalding in water of 70°C or sometimes 95°C. Although early reports by Ward and Lindholm[74] and by Rosenthal et al.[75] had found GF mice significantly less resistant to a 30% 70°C burn than CV mice, a later study by Markley et al.[76] found the opposite. Using ether instead of fluothane as an anesthetic, and with further improvements in the "burn technique", they reported that mortality during the first 48 h after both a 70°C and a 95°C burn was significantly less in GF than in CV mice. Late mortality (after 48 h) was also less in the GF mice. They saw bacterial infection with possible release of LPS in the bloodstream as a significant but by no means the only cause of death during the shock period following burn trauma.

Because of the apparent cruelty of these studies, not much work has been reported lately. However, a recent study by Ma,[77] reported in the *Chinese Medical Journal*, described the results of "25% to 30% burn injury" in GF mice and GF rats and their CV counterparts. The author concludes that infection of gut origin may play an important role in irreversible burn shock and/or early fulminating septicemia.

Intestinal Strangulation

Bacterial LPS has been regarded as a major cause of death in experimental strangulation obstruction. This is confirmed by the fact that GF rats are more tolerant of intestinal strangulation.[78-80] All workers agree that strangulated GF rats survive two to five times longer than their CV counterparts. Amudsen and Gustafsson[79] then studied the strangulation obstruction fluid and found that it contained a great number of bacteria and possibly other pathogenic materials. Obviously, significant translocation of intestinal microorganisms had occurred. When injected i.p. into mice, this fluid caused death within 24 h.

Whereas intestinal strangulation obstruction will always result in ischemia of at least a portion of the intestine, ischemia without obstruction appeared to be acutely lethal for both GF and CV rats. Carter and Einheber[81] ligated the superior mesenteric artery of GF and CV rats; after permanent occlusion GF rats survived for an average of 4.5 h, but CV rats survived about twice as long. After temporary ligation of 1.5, 2, or 3 h the CV rats again showed varied but longer survival times than their GF counterparts. They ascribe this more lethal outcome in the GF animals to toxic materials of nonbacterial origin, possibly vasoactive peptides, which are released from the ischemic intestine. It could be speculated that in such a case the intestinal microflora may afford a certain amount of protection. This brings to mind an early paper by Gordon,[82] in which he reports on the toxic effect of the cecal supernatants of GF mice and rats, an effect very much reduced in CV animals.

Tourniquet Shock

In tourniquet shock the advantage of the absence of an intestinal microflora again becomes obvious. Markley *et al.*[83] subjected mice to 75 min of bilateral hind leg tourniquet trauma. The data show that after 24 h mortality of the GF mice was 8%, against 71% among the CV animals. At 48 h the numbers were 33 and 77%, respectively. The authors cite a number of rather indirect arguments for their plausible claim that LPS is a main factor involved in tourniquet death.

ULCERATIVE COLITIS

In order to establish the role of bacteria in ulcerative colitis, Bylund-Fellenius et al.[84] induced the disease in GF and CV mice of the NMRI/KI strain available from the Karolinska Institute, Stockholm by administering 2.5 or 5% dextran sodium sulfate with the drinking water. A stable sub-acute colitis with maintained diarrhea and shortening of the colon result-ed, but no mortality or gross rectal bleeding was induced in the CV mice. Under comparable conditions the GF mice developed severe colitis with high mortality. This appears to exclude a critical role of the intestinal microbiota for the induction of this type of colitis.

STUDIES OF SWINE DISEASE

The recent availability of the GF pig has brought extensive application of this animal model for the study of the various microbial diseases that plague the pork industry. Again, the absence of an otherwise undefined microflora makes it possible to create gnotobiotes harboring the offending organism and to study the resulting pathology in a microbiologically controlled environment. Ngeleka et al. studied the disease-causing E. coli strains of the serotype 0115.[85-87] The pathogenic strains contain the O-anti-gen K"V165" and/or the F165(1) fimbrial system, the latter being an important factor in the ability of the bacteria to survive in the blood and spread through the host. The ensuing septicemia appears to be a result of a reduced functional ability of the porcine polymorphonuclear leuko-cytes due to the combined presence of the K"V165" antigen and the F165(1) fimbrial factor.

Dykstra et al.[88] studied another E. coli-related problem: the contribu-tion of Shiga-like toxin I to the disease caused by enterohemorrhagic E. coli. The isolated toxin was injected in GN pigs. The ensuing data indicated that this toxin caused vascular damage and ischemic necrosis in the intestines and brain. Christopher-Hennings et al.[89] infected GN pigs with a verotoxin-producing E. coli serotype 0111:NM (verotoxin 1 posi-tive), resulting in intestinal lesions and diarrhea. They again related this pathology to a toxin-related decrease in immune potential: a lower peri-pheral lymphocyte count, and reduced responses to concanavalin A, phytohemagglutin, and pokeweed mitogens.

Other studies covered the pathology of a number of bacterial and viral diseases. Duimstra et al.[90] used a porcine isolate of enterotoxigenic Bacteroides fragilis to orally infect young GN pigs and studied the resulting pathology. The animals developed diarrhea and became mildly anorexic. Intestinal lesions were characterized by swelling, vacuolation, and exfo-liation of enterocytes, and crypt hyperplasia. When a nonenterogenic strain was administered the isolate colonized the intestinal tract but did not cause intestinal lesions. Orist et al.[91] described the results of oral inoculation of GN and CV pigs with a culture of IS intracellularis (intracellular curved bacteria). In this case, whereas all inoculated CV pigs developed severe lesions of proliferative enteropathy in the ilium, none of the GN pigs developed lesions since the organism appeared not to colonize the intestinal tact in the absence of the usual microflora. Neef et al.[92] have recently detailed the pathology caused by 12 well-defined spirochete strains of porcine origin. They concluded that whereas certain non-Serpulina hyodysenteriae spirochetes are capable of inducing disease in the GN pig model, their role as primary or opportunistic pathogens in CV pigs remains dubious.

To reduce resistance and predispose for infection a preinoculation with Bordetella bronchiseptica has often been used. Ackerman et al.[93] inoculated thus pretreated GN pigs with a live, toxigenic strain of Pasteurella multocida at 7 days of age and studied the resulting atrophic rhinitis. Vecht et al.[94] used Bordetella-pretreated GN pigs to study the pathology caused by various strains of S. suis type 2, positive or negative for either the muramidase-released protein (MRP) or the extracellular factor (EF). Strains of the MRP+ EF+ phenotype induced fever, increased the number of polymorphs in blood, and caused meningoencephalitis, polyserositis, and polyarthritis. MRP+ EF− strains caused fever and minor pathology, whereas MRP− EF− phenotype strains induced no signs of disease. Apparently the MRP factor is essential for the disease to fully develop.

In the field of prevention the development of a useful vaccine remains of primary importance. Miniats et al.[95] have tested aluminum hydroxide-absorbed whole-cell bacterins of three strains of Haemophilus parasuis, V1 and V2 (high virulence) and LV (low virulence). V1- and V2-derived vaccines protected GN pigs against challenge with these strains. LV-derived vaccine protected only against V2, but not against V1. Antibodies detected in the sera of the vaccinated pigs were to the outer membranes of the bacteria, but not against homologous LPS or capsular polysaccharides, suggesting that for GN pigs the outer membranes of these organisms are more immunogenic than LPS or capsular antigens. Fedorka-Cray et al.[96] studied the efficiency of a cell-free concentrate prepared from mid-log phase growth cultures of Actinobacillus pleuropneumoniae serotype 1. The preparations contained carbohydrate, LPS, and protein and showed hemolytic and cytotoxic activity. This time the tests were done with pleuropneumonia-free SPF pigs, but GN pigs were used to test antibody formation against the preparation. Results showed that the acellular vaccine provided complete protection from mortality and significantly reduced morbidity to challenge with A. pleuropneumoniae, while the antibody response produced by the vaccine in GN pigs was identical to the one following live challenge.

Diseases of viral origin create additional problems in the pork industry. Collins et al.[97] described the pathology resulting from infection with the SIRS (swine infertility and respiratory syndrome) virus. They exposed 3-day-old GN piglets intranasally to tissue homogenates of originally infected animals. The piglets became anorexic and febrile after 2 to 4 days and developed interstitial pneumonitis and encephalitis, a pathology comparable to that seen out in the field. After two blind passages in GN piglets their tissue homogenates were cultured on cell line CL2621 and a cytopathic virus, provisionally called SIRS virus (ATCC VR-2332), was isolated. When intranasally introduced into GN piglets these animals developed the same pathology as seen in the GN piglets treated with the original tissue homogenate.

Halbur et al.[98] have described the syndrome caused by porcine respiratory coronavirus (PRCV), characterized by necrotizing and proliferative bronchointerstitial pneumonia. Microscopic lesions were mild 3 days postinoculation, were severe at 10 days, but were mostly resolved at day 15. No lesions were observed in the intestine, and there was no clinical respiratory disease. The virus could be isolated from the lungs and from nasal and rectal swabs. At this time these very recent studies aim only at an understanding of the pathology involved in these viral diseases.

REFERENCES

1. Gordon, H. A., Bruckner-Kasdoss, E., and Wostmann, B. S., Aging in germfree mice: life tables and lesions observed at natural death, *J. Gerontol.*, 21, 380, 1966.
2. Walburg, H. E. and Cosgrove, G. E., Aging in irradiated and unirradiated germfree ICR mice, *Exp. Gerontol.*, 2, 143, 1967.
3. Kellogg, T. F. and Wostmann, B. S., Stock diet for colony production of germfree rats and mice, *Lab. Anim. Care*, 19, 812, 1969.
4. Deerberg, F. and Kaspareit, J., Endometrial carcinoma in BDII/Han rats: model of a spontaneous hormone-dependent tumor, *J. Natl. Cancer Inst.*, 78, 1245, 1987.
5. McCay, C., Cromwell, M., and Maynard, L., The effects of retarded growth upon the length of life and upon ultimate size, *J. Nutr.*, 10, 63, 1935.
6. Van der Rijst, M. P., Jansen, B. C., Beeker, T. W., and Wostmann, B. S., Experiments to determine the nutritive value of the average diet consumed in the Netherlands, when fed to white rats *ad libitum* and under conditions of restricted consumption (70%), *Voeding*, 16, 708, 1955.
7. Masoro, E. J., Yu, B. P., and Bertrand, H. A., Action of food restriction in delaying the aging process, *Proc. Natl. Acad. Sci. U.S.A.*, 79, 4239, 1982.
8. Snyder, D. L. and Wostmann, B. S., The design of the Lobund Aging Study and the growth and survival of the Lobund-Wistar rat, in *Dietary Restriction and Aging*, Snyder, D. L., Ed., Alan R. Liss, New York, 1989, 39.
9. Snyder, D. L. and Wostmann, B. S., *Dietary Restriction and Aging*, Snyder, D. L., Ed., Alan R. Liss, New York, 1989.
10. Laquer, G. L., Carcinogenic effects of cycad meal and cycasin, methylazoxymethanol glcoside, in rats and effects of cycasin in germfree rats, *Fed. Proc.*, 23, 1386, 1964.
11. Spatz, M., Smith, D. W. E., McDaniel, E. G., and Laquer, G. L., Role of intestinal microorganisms in determining cycasin toxicity, *Proc. Soc. Exp. Biol. Med.*, 124, 691, 1967.

12. Laqueur, G. L., McDaniel, E. G., and Matsumoto, H., Tumor induction in germfree rats with methylazoxymethanol (MAM) and synthetic MAM acetate, *J. Natl. Cancer Inst.*, 39, 355, 1967.

13. Iwasaki, I., Yumoto, N., Iwase, H., and Ide, G., Potentiation of large intesine tumorigenicity of cycasin derivative by high-fat diet and lactobacillus in germfree mice, *Acta Pathol. Jpn.*, 33, 1197, 1983.

14. Weisburger, J. H., Reddy, B. S., Narisawa, T., and Wynder, E., Germfree status and colon tumor induction by N-methyl-N'-nitro-N-nitrosoguanidine, *Proc. Soc. Exp. Biol. Med.*, 148, 1119, 1975.

15. Balish, E., Shih, C. N., Croft, W. A., Pamukcu, A. M., Lower, G., Bryan, G. T., and Yale, C. E., Effect of age, sex, and intestinal flora on the induction of colon tumors in rats, *J. Natl. Cancer Inst.*, 58, 1103, 1977.

16. Reddy, B. S., Narasawa, T., Weisburger, J. H., and Wynder, E. L., Promoting effect of sodium deoxycholate on colon adenocarcinomas in germfree rats, *J. Natl. Cancer Inst.*, 56, 441, 1978.

17. Weisburger, J. H., Grantham, P. H., Horton, R. E., and Weisburger, E. K., Metabolism of the carcinogen N-hydroxy-N-2-fluorenylacetamide, *Biochem. Pharmacol.*, 19, 151, 1970.

18. El Bayoumy, K., Sharma, C., Louis, Y. M., Reddy, B., and Hecht, S. S., The role of intestinal microflora in the metabolic reduction of 1-nitropyrene to 1-aminopyrene in conventional and germfree rats and in humans, *Cancer Lett.*, 19, 311, 1983.

19. El Bayoumy, K., Reddy, B., and Hecht, S. S., Identification of ring-oxidized metabolites of 1-nitropyrene in the feces and urine of germfree F344 rats, *Carcinogenesis*, 5, 1371, 1984.

20. Morton, K. C. and Wang, C. Y., Enhanced macromolecular binding of N-[4-(5-nitro-2-furyl)-2-thiazolyl]-formamide in germfree vs. conventional rats, *Cancer Res.*, 43, 3628, 1983.

21. Declos, K. B., Cerniglia, C. E., Dooley, K. L., Campbell, W. L., Franklin, W., and Walker, R. P., The role of the intestinal microflora in the metabolic activation of 6-nitrochrysene to DNA-binding derivatives in mice, *Toxicology*, 60, 137, 1990.

22. Yang, Y., Sjövall, J., Rafter, J., and Gustafsson, J. A., Characterization of neutral metabolites of benzo[a]pyrene in urine from germfree rats, *Carcinogenesis*, 15, 681, 1994.

23. George, S. E., Chadwick, R. W., Kohan, M. J., Allison, J. C., Williams, R. W., and Chang, J., Role of the intestinal microbiota in the activation of the promutagen 2,6-dinitro-toluene to mutagenic urine metabolites and comparison of GI enzyme activities in germfree and conventionalized male Fischer 344 rats, *Cancer Lett.*, 79, 181, 1994.

24. Pollard, M., Spontaneous adenocarcinomas in aged germfree Wistar rats, *J. Natl. Cancer Inst.*, 51, 1235, 1973.

25. Pollard, M. and Luckert, P. H., Promotional effects of testosterone and dietary fat on prostate carcinogenesis in genetically susceptible rats, *Prostate*, 5, 1, 1985.

26. Snyder, D. L., Pollard, M., Wostmann, B. S., and Luckert, P., Life span, morphology, and pathology of diet-restricted germfree and conventional Lobund-Wistar rats, *J. Gerontol.*, 45, B52, 1990.

27. Zimmerman, R. J., Gaillard, E. T., and Goldin, A., Pulmonary tumor colony formation following i.v. inoculation of six human colorectal carcinoma xenografts in young gnotobiotic athymic mice, *Clin. Exp. Metastasis*, 6, 27, 1988.

28. Wostmann, B. S., Beaver, M., Bartizak, K., and Madsen, D., Gnotobiotic gerbils, in *Proceedings of the 4th International Symposium on Contamination Control*, Vol. 4, Washington, D.C., 1978, 132.

29. Bartizal, K. F., Beaver, M. H., and Wostmann, B. S., Cholesterol metabolism in gnotobiotic gerbils, *Lipids*, 17, 791, 1982.

30. Orland, F. J., Blayney, J. R., Harrison, R. W., Reyniers, J. A., Trexler, P. C., Wagner, M., Gordon, H. A., and Luckey, T. D., Use of the germfree animal technic in the study of experimental dental caries, *J. Dent. Res.*, 33, 147, 1954.

31. Orland, F. J., Blayney, J. R., Reyniers, J. A., Trexler, P. C., Ervin, R. F., Gordon, H. A., and Wagner, M., Experimental caries in rats inoculated with Enterococci, *J. Am. Dent. Assoc.*, 50, 259, 1955.

32. Fitzgerald, R. J., Jordan, H. V., and Stanley, H. R., Experimental caries and gingival pathologic changes in the gnotobiotic rat, *J. Dent. Res.*, 39, 925, 1960.

33. Fitzgerald, R. J., Dental caries research in gnotobiotic animals, *Caries Res.*, 2, 139, 1968.

34. Rosen, S., Lenney, W. S., and O'Malley, J. E., Dental caries in gnotobiotic rats inoculated with *Lactobacillus casei*, *J. Dent. Res.*, 47, 358, 1968.

35. Jordan, H. V., Bacteriological aspects of experimental dental caries, *Ann. N.Y. Acad. Sci.*, 131, 905, 1965.

36. Kolenbrander, P. E., Ganeshkumar, N., Cassels, F. J., and Hughes, C. V., Coaggregation: specific adherence among human oral plaque bacteria, *FASEB J.*, 7, 406, 1993.

37. Mikx, F. H. M., van der Hoeven, J. S., König, K. G., Plasschaert, A. J. M., and Guggenheim, B., Establishment of defined microbial ecosystems in germfree rats, *Caries Res.*, 6, 211, 1972.

38. Van der Hoeven, J. S., Mikx, F. H. M., Plasschaert, A. J. M., and König, K. G., Methodological aspects of gnotobiotic caries experimentation, *Caries Res.*, 6, 203, 1972.

39. Van Houte, J. and Russo, J., Variable colonization by oral streptococci in molar fissures of monoinfected gnotobiotic rats, *Infect. Immun.*, 52, 620, 1986.

40. Rosen, S., Comparison of sucrose and glucose in the causation of dental caries in gnotobiotic rats, *Arch. Oral Biol.*, 14, 445, 1969.

41. Wilcox, M. D. P., Drucker, D. B., and Green, R. M., Relative cariogenicity and *in-vivo* plaque-forming ability of the bacterium *Streptococcus oralis* in gnotobiotic WAG/RIJ rats, *Arch. Oral Biol.*, 32, 455, 1987.

42. Wilcox, M. D. P., Drucker, D. B., and Green, R. M., Comparative cariogenicity and dental plaque-forming ability in gnotobiotic rats of four species of mutans streptococci, *Arch. Oral Biol.*, 34, 825, 1989.

43. Harris, G. S., Michalek, S. M., and Curtiss, R., The cloning of a locus involved in *Streptococcus mutans* intracellular polysaccharide accumulation and virulence testing of an intracellular polysaccharide-deficient mutant, *Infect. Immun.*, 60, 3175, 1992.

44. Barletta, R. G., Michalek, S. M., and Curtiss, R., Analysis of the virulence of *Streptococcus mutans* serotype c gtfa mutans in the rat model system, *Infect. Immun.*, 56, 322, 1988.

45. Fitzgerald, R. F., Adams, B. O., Sandham, H. J., and Abhyankar, S., Cariogenicity of a lactate dehydrogenase-deficient mutant of *Streptococcus mutans* serotype c in gnotobiotic rats, *Infect. Immun.*, 57, 823, 1989.

46. Wagner, M., Relationship of specific antibacterial agglutinins in saliva to dental caries in gnotobotic rats, in *Germfree Research. Biological Effects of Gnotobiotic Environments. Proceedings of the IV International Symposium on Germfree Research*, Heneghan, J. B., Ed., Academic Press, New York, 1973, 211.

47. McGhee, J. R., Michalek, S. M., Webb, J. R., Navia, J. M., Rahman, A. F. R., and Legler, D. W., Effective immunity to dental caries: protection of gnotobiotic rats by local immunization with *Streptococcus mutans*, *J. Immunol.*, 114, 300, 1975.

48. Gregory, R. L., Michalek, S. M., Shechmeister, I. L., and McGhee, J. R., Effective immunity to dental caries. Protection of gnotobiotic rats by local immunization with a ribosomal preparation from *Streptococcus mutans*, *Microbiol. Immunol.*, 27, 787, 1983.

49. Baer, P. N. and Newton, W. L., The occurrence of periodontal disease in germfree mice, *J. Dent. Res.*, 33, 1238, 1959.

50. Baer, P. N., Newton, W. L., and White, C. L., Studies on periodontal disease in the mouse, *J. Periodontol.*, 35, 388, 1964.

51. Taubman, M. A., Buckelew, J. M., Ebersole, J. L., and Smith, D. J., Periodontal bone loss and immune response to ovalbumin in germfree rats fed antigen-free diet with ovalbumin, *Infect. Immun.*, 32, 145, 1981.

52. Reyniers, J. A., Trexler, P. C., Scruggs, W., Wagner, M., and Gordon, H. A., Observations on germfree and conventional albino rats after total-body X-radiation, *Radiat. Res.*, 5, 591, 1956.

53. McLaughlin, M. M., Dacquisto, M. P., Jacobus, D. P., and Horowitz, R. E., Effects of the germfree state on responses of mice to whole-body irradiation, *Radiat. Res.*, 23, 333, 1964.

54. Onoue, M., Uchida, K., Yokokura, T., Takahashi, T., and Mutai, M., Effect of the intestinal microflora on the survival time of mice exposed to lethal whole-body gamma irradiation, *Radiat. Res.*, 88, 553, 1981.

55. van Bekkum, D. W., Radiation biology, in *The Germfree Animal in Research*, Coates, M. E. *et al.*, Eds., Academic Press, London, 1968, 237.

56. Walburg, H. E., Mynatt, E. I., and Robie, D. M., The effect of strain and diet on the thirty-day mortality of X-irradiated germfree mice, *Radiat. Res.*, 27, 616, 1966.

57. Smith, W. W., Brecher, G., Budd, R. A., and Fred, S., Effects of bacterial endotoxin on the occurrence of spleen colonies in irradiated mice, *Radiat. Res.*, 27 369, 1966.

58. Ledney, G. D. and Wilson, R., Protection induced by bacterial endotoxin against whole-body X-irradiation in germfree and conventional mice, *Proc. Soc. Exp. Biol. Med.*, 118, 1062, 1965.

59. Wilson, R. and Matsuzawa, T., Oxygen changes in tissues of germfree and conventional mice after inoculation of bacterial endotoxin, *Life Sci.*, 7, 1075, 1968.

60. Balish, E., Endotoxin effects in germfree animals, in *Handbook of Endotoxins, Vol. 2: Pathophysiology of Endotoxin*, Hishaw, L. B., Ed., Elsevier Science, Amsterdam, 1985, 338.

61. Heneghan, J. B., Response of germfree animals to shock, *J. Med.*, 21, 51, 1990.

62. Landy, M., Whitby, J. L., Michael, J. G., Woods, M. W., and Newton, W. L., Effect of bacterial endotoxin in germfree mice, *Proc. Soc. Exp. Biol. Med.*, 109, 352, 1962.

63. Jensen, S. B., Mergenhagen, S. E., Fitzgerald, R. J., and Jordan, H. V., Susceptibility of conventional and germfree mice to lethal effects of endotoxin, *Proc. Soc. Exp. Biol. Med.*, 113, 710, 1963.

64. Parant, F., Parant, M., Charlier, H., Sacquet, E., and Chedid, L., Etude de la tolérance aux endotoxines chez la souris "sans germes" au moyen d'un antigéne radioactif, *Ann. Inst. Pasteur*, 110, 198, 1965.

65. Raetz, C. R. H., Ulevitch, R. J., Wright, S. D., Sibley, C. H., Ding, A., and Nathan, C. F., Gram-negative endotoxin: an extraordinary lipid with profound effects on eukaryotic signal transduction, *FASEB J.*, 5, 2652, 1991.

66. Bulanda, M., Zaleska, M., Mandel, L., Talafantova, M., Travnicek, J., Kunstmann, G., Mauff, G., Pulverer, G., and Heczko, P. B., Toxicity of staphylococcal toxic shock syndrome toxin 1 for germfree and conventional piglets, *Rev. Infect. Dis.*, 11, S1, S248, 1989.

67. Zweifach, B. W., Gordon, H. A., Wagner, M., and Reyniers, J. A., Irreversible hemorrhagic shock in germfree rats, *J. Exp. Med.*, 107, 437, 1958.

68. McNulty, W. P. and Linares, R., Hemorrhagic shock of germfree rats, *Am. J. Physiol.*, 198, 141, 1960.

69. Heneghan, J. B., Hemorrhagic shock in unanesthetized gnotobiotic rats, in *Advances in Germfree Research and Gnotobiology. Proceeding of the International Symposium on Germfree Life Research, Nagoya and Inuyama, Japan*, Miyakawa, M. and Luckey, T. D., Eds., CRC Press, Boca Raton, FL, 1967, 166.

70. Yale, C. E. and Torhorst, J. B., Critical bleeding and plasma volumes of the adult germfree rat, *Lab. Anim. Sci.*, 22, 497, 1972.

71. Heneghan, J.B., Hemorrhagic shock in cecectomized germfree rats, in *Germfree Research: Biological Effects of Gnotobiotic Environments. Proceedings of the IV International Symposium on Germfree Research*, Heneghan, J. B., Ed., Academic Press, New York, 1973, 541.

72. Heneghan, J. B. and Stevens, N. C., Cardiovascular response to terminal hemorrhage in gnotobiotic and conventional beagles, *Fed. Proc.*, 34, 380, 1975.

73. Flanagan, J. J., Rush, B. F., Murphy, T. F., Smith, S., Machiedo, G. W., Hsieh, J., Rosa, D. M., and Heneghan, J. B., A "treated" model for severe hemorrhagic shock: a comparison of conventional and germfree animals, *J. Med.*, 21, 104, 1990.

74. Ward, T. G. and Lindholm, L. I., Experimental burns in the germfree animal, *Fed. Proc.*, 19, 103, 1960.
75. Rosenthal, S. R., Ward, T. G., Lindholm, L. I., and Spurrier, W., "Antitoxin" phenomena in burned or injured germfree rats and mice, *Fed. Proc.*, 20, 32, 1961.
76. Markley, K., Smallman, E., Evans, G., and McDaniel, E., Mortality of germfree and conventional mice after thermal trauma, *Am. J. Physiol.*, 209, 365, 1965.
77. Ma, L., Experimental study on the relationship between burn shock and infection, *Chin. Med. J.*, 71, 195, 1991.
78. Cohn, I., Floyd, C. E., Dresden, C. F., and Bornside, G. H., Strangulation obstruction in germfree animals, *Ann. Surg.*, 156, 692, 1962.
79. Amudsen, E. and Gustafsson, B. E., Results of experimental intestinal strangulation obstruction in germfree rats, *J. Exp. Med.*, 117, 823, 1963.
80. Yale, C. E. and Altemeier, W. A., Strangulation obstruction in germfree rats, Surgical Forum, Clin. Congr. 1964, Vol. XV, American College of Surgeons, Chicago, 1964, 294.
81. Carter, D. and Einheber, A., Intestinal ischemic shock in germfree animals, *Surg. Gynecol. Obstet.*, 122, 66, 1966.
82. Gordon, H. A., A bioactive substance in the caecum of germ free animals, *Nature*, 205, 571, 1965.
83. Markley, K., Smallman, E., and Evans, G., Mortality due to endotoxin in germfree and conventional mice after tourniquet trauma, *Am. J. Physiol.*, 219, 541, 1967.
84. Bylund-Fellenius, A. C., Landström, E., Axelsson, L. G., and Midtvedt, T., Experimental colitis induced by dextran sulfate in normal and germfree mice, *Microb. Ecol. Health Dis.*, 7, 207, 1994.
85. Ngeleka, M., Harel, J., Jacques, M., and Fairbrother, J. M., Characterization of a polysaccharide capsular antigen of septicemic *Escherichia coli* 0115:K "V165":F165 and evaluation of its role in pathogenicity, *Infect. Immun.*, 60, 5048, 1992.
86. Ngeleka, M., Jacques, H., Martineau-Doize, B., Daigle, F., Harel, J., and Fairbrother, J. M., Pathogenicity of an *Escherichia coli* 0115:K"V165" mutant negative for F165(1) fimbriae in septicemia in gnotobiotic pigs, *Infect. Immun.*, 61, 836, 1993.
87. Ngeleka, M., Martineau-Doize, B., and Fairbrother, J. M., Septicemia-inducing *Escherichia coli* 0115:K"165F165(1) resists killing by porcine polymorphonuclear leukocytes *in vitro*: role of F165(1) fimbriae and K"V165" O-antigen capsule, *Infect. Immun.*, 162, 398, 1994.
88. Dykstra, S. A., Moxley, R. A., Janke, B. H., Nelson, E. A., and Francis, D. H., Clinical signs and lesions in gnotobiotic pigs inoculated with Shiga-like toxin I from *Escherichia coli*, *Vet. Pathol.*, 30, 410, 1993.
89. Christopher-Hennings, J., Willgohs, J. A., Francis, D. H., Raman, U. A., Moxley, R. A., and Hurley, D. J., Immunocompromise in gnotobiotic pigs induced by verotoxin-producing *Escherichia coli* (0111:NM), *Infect. Immun.*, 61, 2304, 1993.
90. Duimstra, J. R., Myers, L. L., Collins, J. E., Benfield, D. A., Shoop, D. S., and Bradbury, W. C., Enterovirulence of enterotoxigenic *Bacteroides fragilis* in gnotobiotic pigs, *Vet. Pathol.*, 28, 514, 1991.
91. McOrist, S., Jasni, S., Mackie, R. A., McIntyre, N., Neef, N., and Lawson, G. H., Reproduction of porcine proliferative enteropathy with pure cultures of ileal symbiont intracellularis, *Infect. Immun.*, 61, 4286, 1993.
92. Neef, N. A., Lysons, R. J., Trott, D. J., Hampson, D. J., Jones, P. W., and Morgan, J. H., Pathology of porcine intestinal spirochetes in gnotobiotic pigs, *Infect. Immun.*, 62, 2395, 1994.
93. Ackerman, M. R., Rimler, R. B., and Thurston, J. R., Experimental model of atrophic rhinitis in gnotobiotic pigs, *Infect. Immun.*, 59, 3626, 1991.
94. Vecht, U., Wisselink, H. J., van Dijk, J. E., and Smith, H. E., Virulence of *Streptococcus suis* type 2 strains in newborn germfree pigs depends on phenotype, *Infect. Immun.*, 60, 550, 1992.
95. Miniats, O. P., Smart, N. L., and Rosendal, S., Cross protection among *Haemophilus parasuis* strains in immunized gnotobiotic pigs, *Can. J. Vet. Res.*, 55, 37, 1991.

96. Fedorka-Cray, P. J., Stine, D. L., Greenwald, J. M., Gray, J. T., Huether, M. J., and Anderson, G. A., The importance of secreted virulence factors in *Actinobacillus pleuropneumoniae* bacterin preparation: a comparison, *Vet. Microbiol.*, 37, 85, 1993.

97. Collins, J. E., Benfield, D. A., Christianson, W. T., Harris, L., Hennings, J. C., Shaw, D. P., Goyal, S. M., McCullough, S., Morrison, R. B., Joo, H. S., *et al.*, Isolation of swine infertility and respiratory syndrome (isolate ATCC VR-2332) in North America and experimental reproduction of the disease in gnotobiotic pigs, *J. Vet. Diagn. Invest.*, 4, 117, 1992.

98. Halbur, P. G., Paul, P. S., Vaughn, E. M., and Andrews, J. J., Experimental reproduction of pneumonia in gnotobiotic pigs with porcine respiratory coronavirus isolate AR310, *J. Vet. Diagn. Invest.*, 5, 184, 1993.

Chapter XII

APPLICATIONS: PAST, PRESENT, AND FUTURE. PART III. THE PRODUCTION OF MONOCLONAL ANTIBODIES AND CONCLUSIONS

MONOCLONAL ANTIBODIES

The immune system's recognized responsiveness in GF mice maintained on a chemically defined diet (GF-CD mice), coupled with its relatively unstimulated state (see Chapter VII), suggested the use of GF-CD BALB/c mice to effectively generate specific monoclonal antibodies. An antigen injected into such mice would produce a less heterogeneous pool of antibody-producing lymphocytes than it would generate in CV mice maintained on natural ingredient diet L-485 (CV-NI mice), as demonstrated by Bos and Poplis.[1] This would permit enhanced immune response to the less immunogenic regions of an antigen. Taking advantage of this, Gargan *et al.*[2] injected intact cross-linked fibrin into GF-CD mice, and obtained an antibody which recognizes an epitopic region unique to the intact fibrin polymeric structure. The antibody does not cross-react with fibrinogen or with any degradation products of either fibrin or fibrinogen. Its high specificity, coupled to high affinity, makes its monoclonals uniquely suited for localizing at the site of a fibrin clot, or for efficient and safe delivery of coupled clot-dissolving enzymes. Recently, Heidt *et al.*[3] reported the use of anti-interferon-γ monoclonal antibody to control GvH disease in C3H mice that had received a bone marrow graft of C57BL mice. They could mitigate the GvH disease, but could not completely prevent it.

CONCLUSIONS AND FUTURE CONSIDERATIONS

It will be obvious from the size of the last three chapters that over the years the emphasis in the field of gnotobiology has shifted from the study of the GF state and its consequences for function and metabolism, to the use of the gnotobiote for very specific purposes. For this, however, a certain knowledge of that first phase is a necessity for the proper design of present-day experiments, and for the correct interpretation of their results. This is not to say that the first phase is actually over. The first immunological studies actually started in the 1950s, and work with the GF mice maintained on the antigen-free chemically defined diets began in the 1970s and 1980s (Chapters VI and VII). However, especially in the field of immunology, many questions remain, such as the actual nature of "natural antibody"[4] and the ontogeny and function of the present multitude of lymphokines. All these will eventually require absolute control of exogenous antigenicity, which only the inbred GF animal maintained on an antigen-free diet can provide. In the meantime, sufficient knowledge has accumulated to allow the use of the GF and the GN animal in many fields of study.

In the field of nutrition it is mandatory that we come to know what the true systemic requirements are, and to what extent a "normal" microflora might add or subtract, besides its obvious effect on intestinal function. Although the GF rat and the GF mouse have given us models with which to study the later stages of aging without the interference of microbial disease, we yet have to understand why a mild reduction in dietary intake can prolong life substantially. Figure 3 in Chapter II suggests that for the GF rat under the stated conditions the beneficial effect of the absence of a microflora largely disappears. In the field of endocrinology it will be important to further understand the modifications which steroid hormones undergo as they enter the enterohepatic circulation (Chapter II).

For this, as for many other studies, it will be important that we eventually create a gnotobiote which houses a limited, stable, and well-defined microflora of human origin in order to study the many aspects of human disease in which the microflora, directly or indirectly, may play a role. Lately, Rumney et al.[5] have used human-flora-associated rats to study the influence of various human diets on intestinal enzymes, hepatic activation of dietary mutagens, and the formation of hepatic DNA adducts. Results indicated that depending on dietary conditions, the human-derived gut flora either releases a DNA-adducting product able to act outside of the gut or may stimulate the production of such substances otherwise. Thus, under such human-like conditions the formation of DNA-adducting and cross-linking substances appears to take place, and is open for further study. Also, in exploring the use of this "humanized" rat we have to keep the concept of colonization resistance in mind. If we could produce stable and defined microfloras that would protect themselves against the invasion by other especially pathogenic organisms, we might be able to maintain these animals in a clean environment without any further protection. It would be especially important if we could establish such a microflora in the GN gerbil, since this would provide an ideal model for the study of the relationship of cholesterol and bile acid metabolism to cardiovascular disease (see Chapters III and IV).

Then there is the immense and ever more important field of parasitology. Many parasites spend part of their life cycle in the gut of higher organisms. The gut microflora, in creating the environment in which these organisms thrive, must have a substantial influence on their development. Thus far, studies in the U.S., Poland, and Brazil have given us some insight into these matters (Chapter VIII). Obviously though, far more work will be needed before we even begin to understand the role of the intestinal microflora in parasitic diseases. Again, an animal model harboring a stable, human-derived microflora would be of great help in these studies.

The availability of ultrapure chemicals and ultrapure water gives us the possibility to produce an absolutely defined CD diet (Chapter VI). As Dr. Julian Pleasants[6] at the recent meeting of the International Symposium on Gnotobiotics pointed out, GF rats and GF mice maintained on such a diet would make it possible to test very small amounts of materials hitherto considered toxic and/or carcinogenic. This would be in contrast to the backwards extrapolation from high-dose effects presently required by the Delaney Amendment to the U.S. Food, Drug and Cosmetics Act. In this way we might find that very small doses administered in an otherwise totally defined environment may have beneficial instead of adverse effects. In the small doses in which mitogenesis hardly occurs, the induction of systems, enzymatic or immunologic, that can combat or eliminate those materials or their sequelae may far outweigh any deleterious effect.[7]

REFERENCES

1. Bos, N. A. and Poplis, V. A., Humoral immune response to 2,4-dinitrophenyl-keyhole limpet hemocyanin in antigen-free, germfree and conventional BALB/c mice, *Eur. J. Immunol.*, 24, 59, 1994.
2. Gargan, P. E., Gaffney, P. J., Pleasants, J. R., and Poplis, V. A., A monoclonal antibody which recognizes an epitopic region unique to the intact fibrin polymeric structure, *Fibrinolysis*, 7, 275, 1993.
3. Heidt, P. J., Brok, H. P. M., Van der Meide, P. H., Zurcher, C., and Vossen, J. M., Influence of anti-interferon-γ monoclonal antibodies on graft-versus-host disease after allogeneic bone marrow transplantation in mice, in Abstr. 30th Annu. Meet. Assoc. Gnotobiotics, Madison, Wisconsin, 1992, Abstr. #6.
4. Avrameus, S., Natural antibodies: from "horror autotoxicus" to "gnothi seauton", *Immunol. Today*, 12, 154, 1991.

5. Rumney, C. J., Rowland, I. R., Coutts, T. M., Randerath, K., Reddy, R., Shah, A. B., Ellul, A., and O'Neill, I. K., Effects of risk-associated human dietary macrocomponents on processes related to carcinogenesis in human-flora-associated (HFA) rats, *Carcinogenesis*, 14, 79, 1993.
6. Pleasants, J. R., An emerging need for gnotobiotic technology in environmental toxicology, in *Proceedings of the XI International Symposium on Gnotobiology*, Vieira E. and Vieira, L., Eds., (Abstr.).
7. Ames, B. N. and Swirsky Gold, L., Animal cancer tests and cancer prevention, *J. Natl. Cancer Inst. Monogr.*, 12, 125, 1992.

Index

T - #0029 - 160425 - C0 - 234/156/11 [13] - CB - 9780849340086 - Gloss Lamination